T0300116

Green Chemistry, its Role in Achieving Sustainable Development Goals,

Volume 1

Green chemistry practices and principles can play an important role in achieving the United Nations' (UN) sustainable development goals. The expert contributors here have selected key goals and reviewed the implementation of green chemistry for these goals. As described by the UN, it is crucial to harmonize three core elements: economic growth, social inclusion and environmental protection. The sustainable development goals embrace the sustainability mindset, and this will lead to greater productivity and a greener environment. For sustainable development to be achieved, these elements are interconnected and all are crucial for the well-being of individuals and societies.

Features

- The chapters explore sustainable development through green engineering
- Demonstrates the progress made in the search for processes that use fewer toxic chemicals and produce less waste while using less energy
- Highlights the importance of chemistry in everyday life and demonstrates the benefits that the exploitation of green chemistry can have for society
- The pollution of water is of the utmost concern globally, and bioremediation has a strong role to play in ensuring adequate supplies of high-quality water
- These unique volumes address the vast interest in green chemistry and clean processes, which has grown significantly in recent years

Green Chemistry, its Role in Achieving Sustainable Development Goals,

Volume 1

Edited By
Sanjay K. Sharma

CRC Press
Taylor & Francis Group
Boca Raton London New York

CRC Press is an imprint of the
Taylor & Francis Group, an **informa** business

First edition published 2024
by CRC Press
6000 Broken Sound Parkway NW, Suite 300, Boca Raton, FL 33487-2742

and by CRC Press
4 Park Square, Milton Park, Abingdon, Oxon, OX14 4RN

ISBN: 978-1-032-28991-5 (hbk)
ISBN: 978-1-032-29481-0 (pbk)
ISBN: 978-1-003-30176-9 (ebk)

DOI: 10.1201/9781003301769

Typeset in Times
by Deanta Global Publishing Services, Chennai, India

*This book is dedicated to my father the late **Prof. M.P. Sharma** and mother the late **Mrs. Parmeshwari Devi**, the soul inspiration and blessing behind everything I have achieved so far…*

Contents

About the Editor

Dr Sanjay K. Sharma obtained his MSc (1995) and PhD (1999) from the University of Rajasthan, Jaipur, India. He is a Fellow of the prestigious Royal Society of Chemistry, London, UK. He has 25 years of teaching experience at undergraduate and postgraduate levels. Presently, he is working as Professor of Chemistry and Dean (Research & Development) at JECRC University, Jaipur, India. His research interests are Green Chemistry and Water-related research. He has published over 100 research papers in national and international research journals and published 21 books from various international publication houses. He is listed in "World Ranking of Top 2% Scientists" year 2019 (declared in 2020) and 2020 (declared in 2021), published by Stanford University.

Dr. Sharma's title *Bioremediation: A Sustainable Approach to Preserving Earth's Water* is the Choice OAT Award Winner 2021.

Contributors

F.E. Abeng
Department of Chemistry, the Cross River University of Technology, Calabar, Nigeria

M. Açıkyıldız
Ataturk University, Erzurum, Turkey

Swati Agarwal
Drumlins Water Technologies Pvt. Ltd, Jaipur, Rajasthan, 302005, India

Florent Allais
URD Agro-Biotechnologies Industrielles (ABI), CEBB, AgroParisTech, 51110, Pomacle, France

Aastha Anand
Amity International Business School, Amity University, Noida, Uttar Pradesh, India

V.C. Anadebe
Corrosion and Material Protection Division, CSIR-Central Electrochemical Research Institute, Karaikudi 630003, Tami Nadu, India
Academy of Scientific and Innovative Research (AcSIR), Ghaziabad 201002, Uttar Pradesh, India
Department of Chemical Engineering, Alex Ekwueme Federal University, Ndufu Alike, P.M.B. 1010, Abakaliki, Ebonyi State, Nigeria

Süheyda Atalay
Department of Chemical Engineering, Ege University, 35100, Bornova, İzmir, Turkey

Hasan Demir
Department of Chemical Engineering, Osmaniye Korkut Ata University, Osmaniye, Turkey

Gülin Ersöz
Department of Chemical Engineering, Ege University, 35100, Bornova, İzmir, Turkey

Sami Fadlallah
URD Agro-Biotechnologies Industrielles (ABI), CEBB, AgroParisTech, 51110, Pomacle, France

Lei Guo
Department of Material and Chemical Engineering, Tongren University, Tongren 554300, China

A. Gürses
Department of Chemistry Education, Ataturk University, Erzurum, Turkey

K. Güneş
Ataturk University, Erzurum, Turkey

N.B. Iroha
Department of Chemistry, Federal University Otuoke, P.M.B. 126, Yenagoa, Bayelsa State, Nigeria

Suphiya Khan
Department of Bioscience and Biotechnology, Banasthali Vidyapith, Rajasthan, 304022, India

Mahak Kushwaha
Department of Bioscience and
    Biotechnology, Banasthali
    Vidyapith, Rajasthan, 304022,
    India

P.C. Nnaji
Department of Chemical Engineering,
    Michael Okpara University of
    Agriculture, Umudike-P.M.B. 7267
    (Umuahia-Abia State), Nigeria

O.D. Onukwuli
Department of Chemical Engineering,
    Nnamdi Azikiwe University, PMB
    5025, Awka, Anambra State,
    Nigeria

Burcu Palas
Department of Chemical Engineering,
    Ege University, 35100, Bornova,
    İzmir, Turkey

E. Şahin
Ataturk University, Erzurum, Turkey

Meghna Sharma
Economics, Strategy & International
    Business, Amity International
    Business School, Noida, Uttar
    Pradesh, India

Sanjay K. Sharma
Green Chemistry & Sustainability
    Research Group
Department of Chemistry,
    JECRC University,
    Jaipur, Rajasthan, India

Acknowledgments

Writing acknowledgments is an opportunity to express one's gratitude to family, friends, colleagues, and well-wishers for extending their never-ending cooperation and support during the journey of any creation or achievement.

Writing a book on sustainability is no easy task, and I am grateful to all those who have contributed to this effort in one way or another.

First and foremost, I want to thank my family and friends for their unwavering support and encouragement throughout the writing process. Their love and belief in me kept me going even when the going got tough.

I am also indebted to the contributors who have generously shared their insights and experiences with me through their insightful chapters. Their contributions have added depth and richness to the book, and I am honored to have had the opportunity to learn from them. My sincere thanks to them.

I would like to express my gratitude to my publishing team for their expertise and guidance in helping me bring this book to fruition. Their attention to detail and commitment to sustainability has been instrumental in ensuring that this book aligns with its message. I'd like to thank them for providing me with all possible help, assistance, and cooperation during the journey of this book. Working with them is just like working with family. I am happy to have such a fantastic team. Love you guys!

I thank all my students, friends, and colleagues for all their encouragement, cooperation, and support.

Finally, I want to acknowledge the importance of our collective efforts in creating a sustainable future. It is my hope that this book will inspire and empower readers to take action toward a more just and equitable world for all. Thank you all.

Jai Gurudev!

Sanjay K. Sharma, FRSC
7 March 2023
Jaipur, India

Preface

Sustainable growth is assured if ecology is protected. We need to find ways
to maintain harmony in the environment while progressing in science.

—Gurudev Sri Sri Ravishankar

The history of sustainable development goals (SDGs) can be traced back to the
United Nations Conference on Environment and Development (UNCED), held
in Rio de Janeiro, Brazil, in 1992. At this conference, world leaders agreed on
the need to address the interrelated challenges of economic development, social
equity, and environmental protection.

In 2000, the millennium development goals (MDGs) were established as a
set of eight goals to be achieved by 2015, aimed at reducing poverty, improving
health, and promoting sustainable development. The MDGs focused on several
issues, including poverty, education, gender equality, child mortality, maternal
health, infectious diseases, environmental sustainability, and global partnerships.

As the deadline for the MDGs approached, the United Nations began a process of consultations with governments, civil society, and other stakeholders
to develop a new set of sustainable development goals to be implemented from
2015 to 2030. The process culminated in the adoption of the 2030 Agenda for
Sustainable Development by world leaders at the United Nations Sustainable
Development Summit in September 2015.

The 2030 Agenda for Sustainable Development includes 17 SDGs, which are
universal, integrated, and transformative. The SDGs aim to address the root causes
of poverty, inequality, and environmental degradation, and promote sustainable
development in all countries. The SDGs cover a wide range of issues, including
poverty, hunger, health, education, gender equality, clean water and sanitation,
affordable and clean energy, decent work and economic growth, industry, innovation and infrastructure, reduced inequalities, sustainable cities and communities,
responsible consumption and production, climate action, life below water, life on
land, peace, justice, and strong institutions and partnerships for the goals.

In summary, the SDGs are the result of a long process of global consultation
and cooperation, and they represent a shared vision for a sustainable and equitable
future for all.

On the other hand, green chemistry is a field of chemistry that focuses on
designing chemical products and processes that reduce or eliminate the use and
generation of hazardous substances. The relevance of green chemistry lies in its
potential to promote sustainability and address environmental and health concerns associated with traditional chemical processes.

Some of the key benefits of green chemistry include:

1. Environmental protection: Green chemistry aims to minimize the use and release of hazardous chemicals, reducing the environmental impact of chemical processes and products.
2. Improved human health: Green chemistry focuses on reducing the use of toxic and hazardous chemicals, which can help protect workers and consumers from exposure to harmful substances.
3. Economic benefits: Green chemistry can lead to cost savings through the reduction of waste, energy use, and materials consumption.
4. Innovation: The principles of green chemistry promote the development of new, more sustainable technologies and processes, which can spur innovation and create new business opportunities.

In short, green chemistry has the potential to promote sustainable development, protect the environment and human health, and drive economic growth through innovation. And, therefore, adopting the 12 principles of green chemistry can be a useful tool in achieving the SDGs.

Green chemistry can contribute significantly to several of the United Nations' sustainable development goals (SDGs), which aim to create a more sustainable and equitable world. Some of the key ways in which green chemistry can support these goals directly include:

1. Goal 7: Affordable and Clean Energy—Green chemistry can help reduce the carbon footprint of chemical processes by promoting the use of renewable energy sources, such as solar and wind power, and developing more energy-efficient processes.
2. Goal 9: Industry, Innovation, and Infrastructure—Green chemistry encourages the development of new, sustainable technologies and materials that can drive innovation and economic growth while reducing the environmental impact of industry.
3. Goal 11: Sustainable Cities and Communities—Green chemistry can help reduce pollution and emissions in urban areas by promoting the use of sustainable building materials and developing cleaner production methods.
4. Goal 12: Responsible Consumption and Production—Green chemistry promotes the use of renewable feedstocks, the reduction of hazardous substances, and the minimization of waste, all of which support responsible consumption and production.
5. Goal 13: Climate Action—By reducing the carbon footprint of chemical processes, promoting the use of renewable energy, and developing sustainable technologies, green chemistry can help mitigate the impact of climate change.
6. Goal 14: Life Below Water and Goal 15: Life On Land—Green chemistry supports the protection of marine and terrestrial ecosystems by minimizing pollution and reducing the use of hazardous substances.

Therefore, green chemistry has the potential to contribute significantly to the achievement of several sustainable development goals by promoting sustainable and environmentally friendly practices in industry and production.

The present book is a modest effort of presenting the research updates in the area of green chemistry which is directly or indirectly serving the environment and a step ahead towards achieving the SDGs. In this volume, a variety of topics have been covered which are quite useful for students, scholars, researchers, and policymakers equally.

I gratefully acknowledge all the contributors to this book, without whom this book could not have been completed and published. I express my highest gratitude and thankfulness to all of them.

Jai Gurudev!

Sanjay K. Sharma, FRSC
7 March 2023
Jaipur, India

1 Contribution of Sustainable/ Green Chemistry on Sustainable Development Goals
Bibliometric Analysis

Hasan Demir and Sanjay K. Sharma

CONTENTS

1.1 INTRODUCTION

The United Nations (UN) organized the World Summit on Sustainable Development in Johannesburg, South Africa (Report of the UN, 2002). The UN adopted courses of action for the implementation of policies, science, and green technologies in five areas, namely water, energy, health, agriculture, and biodiversity progress. The use of chemicals must be altered to reduce unsustainable consumption and production practices, leading to the reduction of negative impacts on human health and the environment. This can be achieved by managing chemicals throughout their life cycle and hazardous wastes for sustainable development

DOI: 10.1201/9781003301769-1

1

by the year 2020. The UN defined 17 goals for implementing sustainable development, which are as follows:

- Goal 1: No poverty
- Goal 2: Zero hunger through agrochemistry
- Goal 3: Good health and well-being
- Goal 4: Quality education
- Goal 5: Gender equality
- Goal 6: Clean water and sanitation
- Goal 7: Affordable and clean energy
- Goal 8: Decent work and economic growth
- Goal 9: Industry, innovation, and infrastructure
- Goal 10: Reduced inequalities
- Goal 11: Sustainable cities and communities
- Goal 12: Responsible consumption and production
- Goal 13: Climate action
- Goal 14: Life below water
- Goal 15: Life on land
- Goal 16: Peace and justice in strong institutions
- Goal 17: Partnerships (Welton, 2018)

Green and/or sustainable chemistry can contribute directly/indirectly to most of the sustainable development goals (SDGs). In light of the SDGs, scientists and researchers—whether they are employed by academia or industry—began to reconsider, investigate, and discuss how they could rebuild their labs and/or industrial processes to reduce the number of chemicals needed to produce a given effect by a factor of two every five years (Welton, 2018). The current approaches to achieving the SDGs, which are based on the promotion of more environmentally friendly technologies, offer only a limited amount of mitigation and will not be able to support the sustainability of industrialized societies over the long term. For environmental sustainability, the considering of second law of thermodynamics, which provides a model for maintaining a steady state of system, influences delicate balances in biosphere, human activities, including economics and politics, and therefore permanent attention must be paid to that law and the limits it imposes (Marques & Machado, 2014). The use of a less toxic solvent in favor of one that was very hazardous could be claimed as being green. When called "greening" analytical procedures, there are several factors to consider, including the volume of waste produced, the toxicity and environmental impact of all chemicals used, energy, and power utilized (Turner, 2013). Chemical sectors and researchers in respect of green and sustainable chemistry can serve SDGs with the followings:

- Needed a holistic system approach
- Setting policies
- Measurable objectives for a continuous process of improvement

- Increase networking interdisciplinary scientific research
- Promote reuse and recycling
- Education
- Consumer awareness
- More sustainable products in the supply chains
- Corporate social responsibility and sustainable entrepreneurship (Blum et al., 2017)

Scientometrics is a broad area of study that examines how the development of digital information products has led to an enormous amount of scientific literature being published online. The researchers explore, gather, examine, and evaluate this massive amount of data to address the issue of information overload (Derossi et al., 2021). Bibliometrics was identified as "the application of mathematical and statistical methods to books and other communication media" by Pritchard (1969) (Velmurugan & Radhakrishnan 2016). The methods of bibliometrics are used to analyze authorship, citation and publication pattern, and the relationship within scientific domains and research communities and to structures of specific fields (Vijay & Raghavan, 2007; Jermann et al., 2015; Verma et al., 2015; Demir & Sharma, 2019). The analysis of papers with various statistical methods can be directly related to investigation studies, features, and behaviors of published pieces of knowledge for investigation of the structures of research and scientific areas, and assessment of the administration of scientific information and research activity (Velmurugan & Radhakrishnan, 2016).

Bibliometrics offers bibliometric information and high-quality, normalized data in terms of researchers, research centers, or universities as a whole; graphical representation of the data; data extraction in different formats from different indexing databases and faceted search (Rosa et al., 2016). Bibliometrics provides four levels of indicators, which are productivity (year-wise publication, author, and type of publication), visibility (indexing in the Web of Science (WoS), SCOPUS, dimensions, etc.), impact (the number of citations received by year), and collaboration (level of coauthoring of publications to be analyzed in terms of international, national, inter-university, or without collaboration) (Rosa et al., 2016).

This study will survey published articles on the contribution of green and/or sustainable chemistry to sustainable development goals. In this respect, the bibliometric study will be made on papers indexing in Science Citation Index, Science Citation Index Expanded, Journal Citation Reports, and Engineering Index from SCOPUS and WoS databases. The analysis involves mainly a discussion of papers as follows:

- Distribution of papers on year perspective
- Geographical-wise distribution of articles
- Authorship pattern and year perspectives
- Degree of authorship collaboration
- National and international collaboration

- Trends of science categories
- Degree of funding
- Citation

1.2 METHODOLOGY

The bibliometrics search was made on August 2022. The outcome of the search was 2,591 articles published in different journals. These papers were examined in Bibliometrix (Aria & Cuccurullo, 2017) and the MS Excel program.

Figure 1.1 presents the distribution of types of papers in SCOPUS and WoS indexing databases. As indicated by the abovementioned number, SCOPUS and WoS had differing counts of published publications. The WoS Core Collection has 76 million records including 21,000 journals, 111,000 books, and almost 8 million conference papers as of June 2020 (Stahlschmidt & Stephen, 2020). On

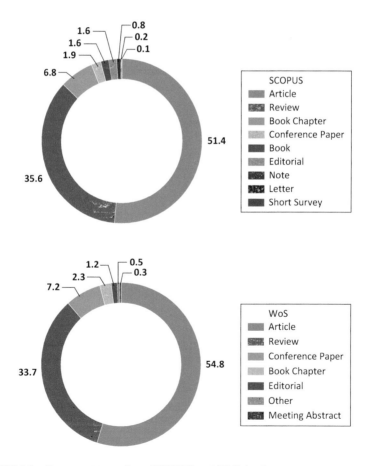

FIGURE 1.1   Document types from SCOPUS and WoS databases.

the SCOPUS side, it indexed almost 78 million items including 23,400 journals, 850 book series, almost 10 million conference papers, and 44 million patents (Stahlschmidt & Stephen, 2020). According to statistics, the fundamental distinction between WoS and SCOPUS is the size and coverage of the databases. Figure 1.1 shows the validation of the statistics, and the percentages of articles and review papers were similar in SCOPUS and WoS. The percentages of book chapters and conference articles showed a substantial difference. In SCOPUS, book chapters and conference articles made up 6.8% and 1.9% of the total, compared to 2.3% and 7.2% in the WoS database. The bibliometric data of the SCOPUS and WoS databases also show differences. Because of this, some of the figures in the remaining sections of this chapter include data from both databases, while others only include data from the WoS or SCOPUS databases.

1.3 GENERAL INFORMATION AND YEAR-WISE DISTRIBUTIONS

Figure 1.2 presents the distribution of published papers according to the years for both indexing in SCOPUS and WoS. The WoS list starts from 1998 and the SCOPUS list begins from 2013. For both databases, the number of papers increased gradually each year. In 2021, the maximum number of papers was 266 for WoS and 393 for SCOPUS. The average published papers per year was 60 and 114 for WoS and SCOPUS, respectively. We might conclude that individuals such as researchers, scientists, and manufacturers who were active between the late 1990s and 2020s recognized the significance of green chemistry on SDGs.

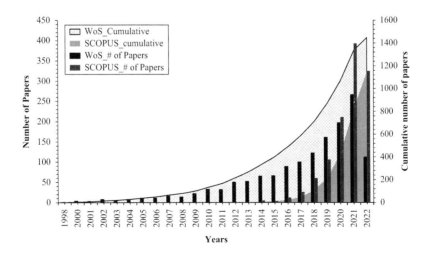

FIGURE 1.2 Year-wise distribution of the number of articles from SCOPUS and WoS databases.

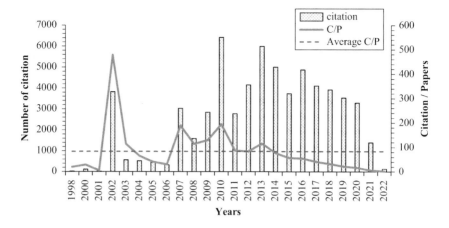

FIGURE 1.3   Year-wise citation score of papers from the WoS database.

1.4   THE FREQUENCY OF CITATIONS

The distribution of the year-wise citation score of papers is shown in Figure 1.3. The maximum citation scores were observed in 2010 and 2013 as 6,421 and 5,991 citations, respectively. The total number of citations during these years was 62,660. The citation score is current data, and it takes time to fully understand the cailber of papers due to newly published research that mentions previous works. The cumulative citation score increases over the years. Figure 1.3 also presents the distribution of the number of citations per paper (C/P) and the ratio of C/P per citable years. Citable years refer to the number of years during which a paper can potentially receive citations after its publication. For instance, if a paper was published in 1998 and 24 years have passed since then, it still has the opportunity to receive citations up to the present time. The average citation per paper was 83 and the average citation per year was 2,610. Papers published in 2002 had the highest number of citations per paper with 478. Papers published in the last eight years have a lower number of C/P than the average C/P. However, recently published papers have a high ratio of C/P per year, which means that they obtained a high number of citations in a short time. Recent research papers show promising results, suggesting that they have the potential to receive a greater number of citations than older papers.

1.5   AUTHORSHIP PATTERN

Figure 1.4 represents the authorship pattern of papers. A total of 6,221 authors, some of whom may have participated in multiple studies, published a total of 1,448 papers that are exclusively indexed in WoS. The exact number of authors was 974 without any repetition. The ratio of paper per author was 1.49; 174 papers were published by 128 single authors. The percentages of papers authored by two, three, and four authors were 18%, 18%, and 15%, respectively. Of these, 1,274

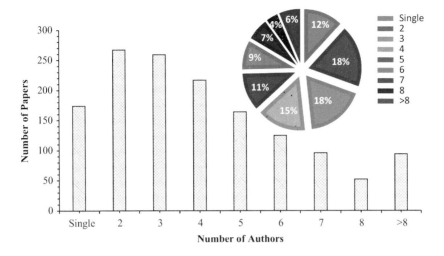

FIGURE 1.4   Authorship pattern of papers from WoS.

papers were written by a collaboration of 846 authors. A ratio of authors per paper can give a collaboration index, which was 1.51 in this case. Additionally, 88% of papers were studied by multiple authors.

Figure 1.5 illustrates the year-wise collaboration index distribution and the variation of authorship of papers against years. The C indicates a collaboration degree and was calculated as the sum of single and multiple authors (Subramanyam, 1983; Demir & Sharma, 2021). The collaboration index fluctuated before 2010 due to the low number of papers. After 2010, the collaboration index was almost steady and reached between 1.1 and 1.3.

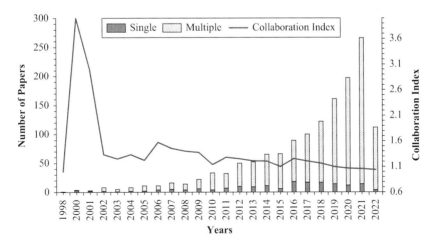

FIGURE 1.5   Year-wise collaboration index and authorship of papers from the WoS database.

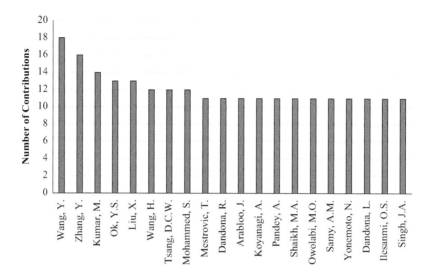

FIGURE 1.6　Top 20 authors' production against years from the SCOPUS database.

Figure 1.6 illustrates the top 20 authors' production from the SCOPUS database over time. Wang Y. published 18 papers between 2014 and 2019 years. This is followed by the author Zhang Y. who published 16 papers and then Kumar M. who published 14 papers. The authors' collaboration network from the SCOPUS database is illustrated schematically in Figure 1.7. In the SCOPUS database, there were 4,611 authors in total. Figure 1.7 was created by selecting a minimum of 4

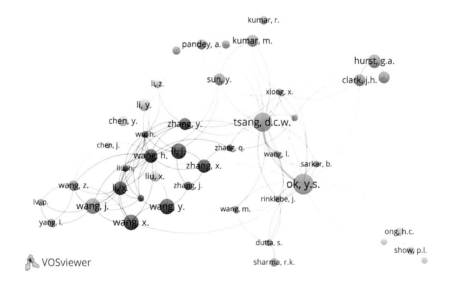

FIGURE 1.7　Authors collaboration network from the SCOPUS database.

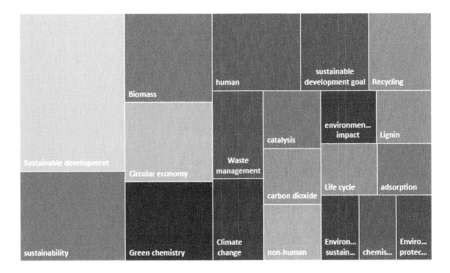

FIGURE 1.8 Trends of science categories of published papers.

documents per author, resulting in 67 papers meeting these criteria. The figure depicts authors who have collaborated with each other, with the most collaborative authors shown as bigger spots using VOSviewer.

1.6 SCIENTIFIC CATEGORIES AND TREND TOPICS

Figure 1.8 indicates trend topics as well as scientific categories of published papers obtained from WoS and SCOPUS databases. The keywords of 2,591 published papers were analyzed and words that were repeated at least 20 times were shown in a treemap graph. This analysis can give information about the contribution of green chemistry to SDGs. The 12 principles of green chemistry show an agreement and/or contribute to the 17 SDGs. The words "sustainability" and "sustainable developments" obviously featured as keywords of published papers. "Human", "non-human", "biomass", "waste management", "recycling", and "climate change" are the supporting keywords for the contribution of green chemistry.

1.7 TOP JOURNALS

Figure 1.9 depicts the name of the top 20 journals that published papers related to topics from the WoS database. Of all papers, 78 (8.4%), were published in the *Green Chemistry* journal. The *ACS Sustainable Chemistry and Engineering* journal published 64 papers (4.4%). The *Current Opinion in Green and Sustainable Chemistry* journal published 43 papers (2.97%).

WoS

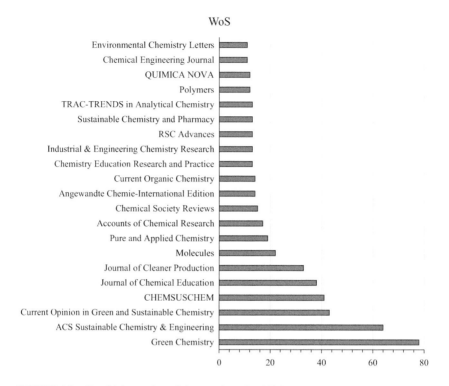

FIGURE 1.9   Top 20 journals and the number of published papers.

1.8   COUNTRY-WISE DISTRIBUTION OF CONTRIBUTIONS AND AFFILIATIONS

Figure 1.10 illustrates the contributions made by countries. The maximum contributions were made by the USA (168). Following the USA, the contributions of China, the UK, and India were 163, 160, and 139, respectively. The 126 countries in the world contributed these 1,591 papers in different journals.

The geographical-wise distribution of contributions was represented on the Earth map as shown in Figure 1.11. The intensity of published research and review papers was shown from green to red color. 126 countries were spread around the six continents. The USA, China, and India become prominent with green color indicating maximum contributions. In terms of international collaborations, the number of national and international contributions to papers is shown in Figure 1.12 from the WoS database. Most of the published papers (60.7%) were published with national and international collaboration. About 39.3% of papers were published with a single address, which means they were published without collaboration. After 2006, the percentage of collaboration increased over the years. The topics were suitable for multidisciplinary collaborations. The percentage of national collaboration (34.1%) was higher than that of international

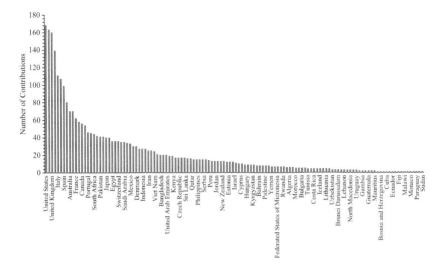

FIGURE 1.10   Country-wise distribution of contributions from the SCOPUS database.

collaboration, which was 26.7%. The number of international collaborations has steadily increased over the past five years because of the internet, improved electronic communication methods, and easier access to literature.

Figure 1.13 depicts the name of the top 20 Institutions that published papers on the contribution of green chemistry to SDGs. The University of York researched 31 studies. The Ministry of Education of China, the Imperial College of London, and the Chinese Academy of Sciences each published 25 papers. The top 20 institutions published 405 papers, which is 35.4% of published papers in the SCOPUS database.

1.9   RESEARCH FUNDING AND SUPPORTERS

Figure 1.14 shows the year-wise funded and non-funded number of papers and the percentage of funding. However, there is no need to emphasize that financial supports for a project are very important for research activities. Especially, the cost of analytical equipment, consumable prices, and labor costs are increasing these days. 877 papers (60.6%) were financially supported, and 571 papers (39.4%) were not supported. As is seen in Figure 1.14, the ratio between funded and non-funded papers fluctuated over the years. However, the percentage of funded papers increases over the years.

The project grants were supplied by national and international organizations, some of which are shown in Figure 1.15. 35.4% of papers were supported by listed funding organizations shown in Figure 1.15 according to the SCOPUS database. The European Commission and the National Natural Science Foundation of China supported 66 and 56 papers, respectively. The Horizon 2020 framework

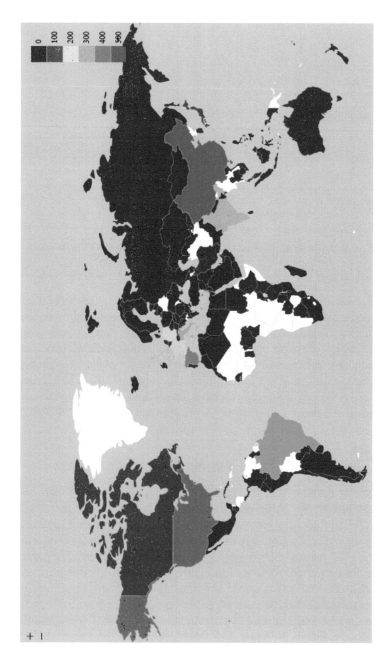

FIGURE 1.11   Geographical-wise distribution of contributions on the Earth map from the WoS database.

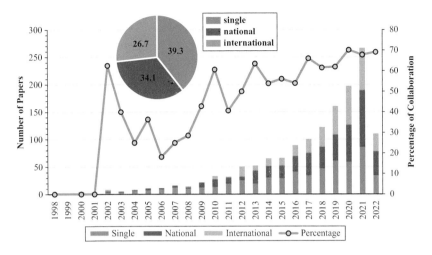

FIGURE 1.12 National and international collaboration with respect to years from the WoS database.

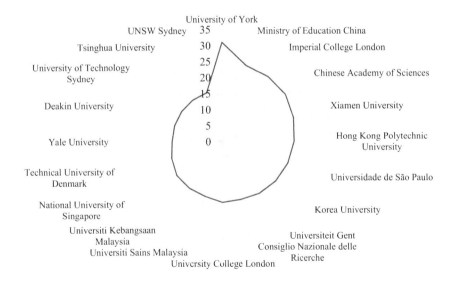

FIGURE 1.13 Top 20 affiliations that most published papers from the SCOPUS database.

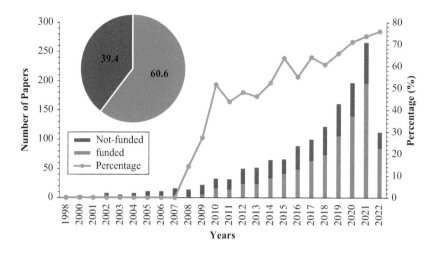

FIGURE 1.14   Year-wise funded and non-funded number of papers and percentage of funding from the WoS database.

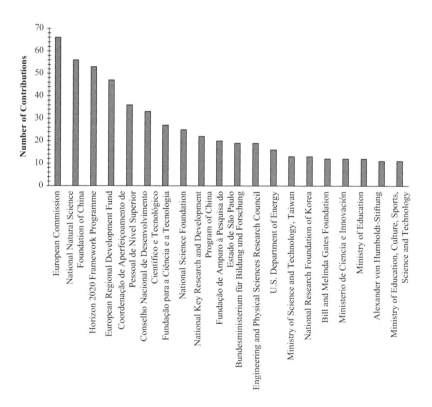

FIGURE 1.15   Top 20 of foundation organizations from the SCOPUS database.

and the European Regional Development Fund were acknowledged 53 and 47 times in the papers, respectively.

1.10   REMARKS

The 2,591 papers indexing in SCOPUS and WoS were extracted and examined in Bibliometrix, MS Excel, and VOSviewer programs. The bibliometric analysis was found out for investigating authorship patterns, international collaboration, and funded/non-funded research. The following remarks were summarized:

- According to the statistics, the fundamental distinction between WoS and SCOPUS is the size and coverage of the databases.
- 6,221 authors from 126 countries contributed concerning the contribution of green chemistry to SDG.
- The exact number of authors was 974 without any repetition. The ratio of paper per author was 1.49.
- 1,274 papers were written by a collaboration of 846 authors. 88% of papers were studied and published by multiple authors. A ratio of authors per paper can give a collaboration index, which was 1.51.
- The total number of citations from 1998 to 2022 was 62,660. The average ratio of C/P per citable years was found to be 83 and the average citations per year were 2,610.
- The scientific categories revealed that 12 principles of green chemistry make contributions to 17 goals of SDGs.
- With 168 papers, the USA contributed the most.
- The 39.3% of papers were published with a single address, and 60.7% of papers were published as national or international collaborations.
- 571 papers (39.4%) were not financially funded, while 877 articles (60.6%) were.
- The European Commission and China's National Natural Science Foundation, which have sponsored 66 and 56 papers, respectively, each provided financing for 10.7% of the papers.

REFERENCES

Aria, M., & Cuccurullo, C. (2017). Bibliometrix: An R-tool for comprehensive science mapping analysis. *Journal of Informetrics, 11*(4), 959–975. https://doi.org/10.1016/j.joi.2017.08.007

Blum, C., Bunke, D., Hungsberg, M., Roelofs, E., Joas, A., Joas, R., Blepp, M., & Stolzenberg, H.-C. (2017). The concept of sustainable chemistry: Key drivers for the transition towards sustainable development. *Sustainable Chemistry and Pharmacy, 5*, 94–104. https://doi.org/10.1016/j.scp.2017.01.001

Charlotta, T. (2013). Sustainable analytical chemistry—More than just being green. *Pure and Applied Chemistry, 85*(12), 2217–2229. http://doi.org/10.1351/PAC-CON-13-02-05

Demir, H., & Sharma, S. K. (2019). Chapter 2: share of bioremediation in research journals A bibliometric study. In S. K. Sharma (Ed.), *Bioremediation: A sustainable approach to preserving earth's water* (pp. 19–34), CRC Press.

Demir, H., & Sharma, S. K. (2021). Chapter 1: Green chemistry and water remediation: Bibliometric study and research applications. In S. K. Sharma (Ed.), *Advances in green and sustainable chemistry* (pp. 1–33), Elsevier.

Derossi, R., Caporizzi, R., Paolillo, M., Oral, M. O., & Severini, C. (2021). Drawing the scientific landscape of 3D food printing; maps and interpretation of the global information in the first 13 years of detailed experiments, from 2007 to 2020. *Innovative Food Science and Emerging Technologies, 70*(June), 102689. https://doi.org/10.1016/j.ifset.2021.102689

Jermann, C., Koutchma, T., Margas, E., Leadley, C., & Ros-Polski, V. (2015). Mapping trends in novel and emerging food processing technologies around the World. *Innovative Food Science and Emerging Technologies, 31*, 14–27.

Marques, C. A., & Machado, A. A. S. C. (2014). Environmental sustainability: Implications and limitations to green chemistry. *Foundations of Chemistry, 16*(2), 125–147. https://doi.org/10.1007/s10698-013-9189-x

Padrós-Cuxart, R., Riera-Quintero, C., & March-Mir, F. Bibliometrics: A publication analysis tool, BIR 2016 workshop on bibliometric-enhanced information Retrieval.

Stahlschmidt Stephan and Stephen Dimity, comparison of Web of Science, Scopus and dimensions databases KB Forschungspoolprojekt (2020). German centre for higher education research and science studies (DZHW) GmbH |www. dzhw.eu October 2020.

Pritchard, A. (1969). Statistical bibliography or bibliometrics. *Journal of Documentation, 25*, 348–349.

Report of the United Nations, report of the world summit on sustainable development Johannesburg, South Africa, 26 August–4 September 2002, New York, 2002.

Subramanyam, K. (1983). Bibliometric studies of research collaboration: A review. *Journal of Information Science, 6*(1), 33–38. https://doi.org/10.1177/016555158300600105

Tom, W. (2018). Editorial overview: Unsustainable development goals: How can sustainable/green chemistry contribute? There can be more than one approach. *Current Opinion in Green and Sustainable Chemistry, 13*, A7–A9. https://doi.org/10.1016/j.cogsc.2018.09.005

Velmurugan, C., & Radhakrishnan, N. (2016). Indian Journal of Biotechnology: A bibliometric study. *Innovare Journal of Science, 4*, 1–7.

Verma, A., Sonkar, S. K., & Gupta, V. (2015). A bibliometric study of the library philosophy and practice (e-journal) for the period 2005–2014. *Library Philosophy and Practice, 1292*. Retrieved January 19, 2018, from http://digitalcommons.unl.edu/libphilprac/1292

Vijay, K. R., & Raghavan, I. (2007). Journal of food science and technology: A bibliometric study. *Annals of Library and Information Studies, 54*, 207–212.

2 The Use of Biomass for the Enhancement of Biogas Production

F.E. Abeng, N.B. Iroha, V.C. Anadebe and Lei Guo

CONTENTS

DOI: 10.1201/9781003301769-2

2.1   INTRODUCTION: BACKGROUND

The United Nations approved sustainable development goals (SDGs) to address environmental sustainability, economic sustainability and social sustainability. The United Nations promotes the use of natural resources like water, energy and raw materials in a sustainable and eco-friendly manner (Alayi et al., 2016). The Paris Climate Change Agreement (PACC) and the 2030 Agenda for Sustainable Development have established 17 sustainable development goals (SDGs) as a shared platform and blueprint for peace and prosperity for humanity and the planet Earth. The concentration of greenhouse gases (GHGs) in the atmosphere has come from fast population growth and industrial advancement (Alayi et al., 2016; Abdelkareem et al., 2021). The Intergovernmental Panel on Climate Change (IPCC) has set a limit of 1.5°C on temperature rises to prevent negative effects on habitats and communities. To combat climate change, it is critical to quickly convert fossil fuels into clean, sustainable and renewable energy sources. Following decades of research and industrial action, the general consensus today is to apply green chemistry or sustainable chemistry in converting waste to energy, which is a viable choice for waste management, as it delivers various benefits (Obaideen et al., 2022). In certain situations, waste can be regarded as a semi-renewable energy source and a substitute for fossil fuels. The green chemistry approach has been developed in governmental, industrial and educational fields worldwide to reduce the use of chemical reagents, waste products and hazardous substances during the production of fossil fuels (Hofer & Bigorra, 2008). The biogas industry is uniquely positioned to help achieve nine of the SDGs which include industry, innovation and infrastructure, generating clean energy, mitigating the effects of climate change, reducing poverty and delivering social justice – perhaps more than any other sector (Akkarawatkhoosith et al., 2019). Biogas is involved in many different parts of the economy and environment; this is because there are lots of biodegradable wastes around us, which are derived from multitudes of human, social and economic activities. Such wastes can be found in food waste from homes, restaurants, shops and caterers; waste from industrial production; and agricultural wastes from animal husbandry, crop cultivation and food production, which we call biomass (Akkarawatkhoosith et al., 2019; Mishra et al., 2021). All of these wastes produce methane (CH_4), which may be collected and transported to anaerobic digestion (AD) plants to generate sustainable heat and electricity for local use or distribution into larger grids (Herwintono et al., 2020). Anaerobic digestion (AD) is a natural process in which bacteria digest organic matter (OM) in enclosed containers to produce biogas, which can be used for cooking, heating, cooling and electricity generation, or improved and utilized as automobile fuel or gas grid injection (Herwintono et al., 2020).

2.2   BIOGAS

Biogas is a mixture of gases made from agricultural waste, manure, municipal trash, plant material, sewage, green waste and food waste, with the main components being methane, carbon dioxide and hydrogen sulphide (Bremond et al., 2021). Anaerobic digestion with anaerobic organisms or methanogen in an anaerobic digester, biodigester or bioreactor produces biogas. Biogas is largely composed of methane (CH_4) and carbon dioxide (CO_2), with minor amounts of hydrogen sulphide (H_2S), moisture and siloxanes present. Methane, hydrogen and carbon monoxide (CO) are all gases that can be burned or oxidized with oxygen (Shaibur et al., 2021). Biogas can be utilized as a fuel because of the energy release; it can be used in fuel cells and for any heating purpose, such as cooking. It can also be utilized to transform the energy in gases into electricity and heat in a gas engine. Biogas can be compressed and utilized to power motor vehicles after carbon dioxide and hydrogen sulphide are removed (Lundmark et al., 2021). Biogas, for example, has the potential to replace about 17% of car fuel in the United Kingdom. When biogas is converted to biomethane, it can be cleaned and upgraded to natural gas standards. Biogas is a renewable resource since it has a continual production and usage cycle and produces no net carbon dioxide. Microorganisms that undergo anaerobic respiration, such as methanogens and sulphate-reducing bacteria, produce biogas. Biogas can be applied to both natural and industrially produced gas (Xue et al., 2020).

2.3   CLASSIFICATION OF ENERGY SOURCE

There are two major types of energy sources: renewable and non-renewable. Solar, wind, hydro, marine, geothermal, biomass and other sources of energy that are produced from "solar energy", and that can thus be renewed forever in nature, are all examples of renewable energy (Khoiyangbam et al., 2011; Abdelkareem et al., 2021). Renewable resources are inexhaustible resources that can be recycled, reproduced, regenerated or replaced. Within a few human generations, renewable resources can be replenished. Because some resources can be replaced over very long geological timescales, the word "few generations" is crucial. For example, rocks are recycled thousands of times slower in nature than they are utilized, making them nearly non-renewable (Tansel & Surita, 2019). Non-renewable energy resources are those that are replenished by extremely slow natural cycles or are not recycled at all for practical purposes. Solar energy is seen as renewable energy despite the sun's limited lifespan (Zhang et al., 2019). Biogas is an energy source that is renewable. The energy for biogas production originates from the Sun, through plant photosynthesis. The solar energy-storing plant biomass is used to power the biogas plant either directly or after partial digestion in animal stomachs. Plant biomass, both dry and green, is consumed by ruminants (Khoiyangbam et al., 2011). Crop straw, cereal residues, pulses and oilseeds are commonly consumed as dry fodder, while grasses from permanent pastures and forest areas are typically fed to these animals in mountain locations. Biogas is produced from animal waste that has undergone anaerobic digestion in digesters. The complex organic polymers in the biomass, predominantly carbohydrates, lipids and proteins, are fermented inside the biogas plant to produce biogas, which is mostly made up of methane and carbon dioxide. Biogas has gained significant traction as a viable alternative to traditional energy sources in recent decades, particularly in emerging nations such as China and India (Khoiyangbam et al., 2011). It is obvious that traditional energy sources pollute the environment and harm human health. Furthermore, because fossil fuels are costly and limited, power generation based on them cannot be sustained over time. This emphasizes the importance of investigating the potential for renewable fuel replacement and the use of fuel-efficient devices. Renewable energy has a number of advantages, including (Khoiyangbam et al., 2011):

(1) Its perennial nature
(2) The use of locally available resources that do not require elaborate transportation arrangements
(3) Suitability for decentralized applications and use in remote areas
(4) Low gestation and less capital-intensive nature
(5) Modular nature, which means small-scale units and systems can be almost as cost-effective as large-scale ones
(6) Environmental friendliness
(7) Effective use for both augmenting the availability of power and as a tool for rural development and social justice

As a result, there is a rising global consensus in favour of renewable energy sources as clean and sustainable energy sources.

2.4 PRINCIPLE OF ANAEROBIC DIGESTION FOR BIOGAS PRODUCTION

Anaerobic digestion (AD) is a waste-to-biogas conversion process. AD is where organic wastes are reintegrated into ecosystem dynamics by a natural method on earth. This process has been studied for many years and is now well known as an active topic in science (Hofer & Bigorra, 2008; Wang et al., 2020). AD is produced by multiple interdependent microorganism communities that live in an oxygen-free environment and that transfer complex substrates in four stages: hydrolysis, acetogenesis, acidogenesis and methanogenesis are all examples of reactions that occur in anaerobic digestion (AD). Biogas via anaerobic digestion (AD) is an energy-efficient and environmentally friendly technology that offers significant advantages over other types of bioenergy (Hofer & Bigorra, 2008; Calbry-Muzyka et al., 2019). By utilizing locally available sources, anaerobic digestion (AD) technology significantly reduces greenhouse gas (GHG) emissions when compared to fossil fuels. Furthermore, digestate, a by-product used as fertilizer, has a high value in crop growth and can effectively substitute common mineral fertilizers (Hofer & Bigorra, 2008). The reactions that occur in anaerobic digestion (AD) are discussed in the following.

2.4.1 Hydrolysis

Hydrolysis is a water-based reaction. Figure 2.3 depicts the hydrolysis reaction and how water and enzymes can break down cellulose, starch and simple sugars according to Equation (2.1). In anaerobic digestion, the enzymes are exoenzymes (cellulosome, protease, etc.) from a number of bacteria, protozoa and fungi (Bharathiraja et al., 2018; Elijah et al., 2009; Zhang et al., 2019).

$$\text{Biomass} + H_2O \rightarrow \text{Monomers} + H_2. \qquad (2.1)$$

2.4.2 Acidogenesis

Soluble monomers are transformed during acidogenesis into tiny organic molecules such as short-chain (volatile) acids (propionic, formic, lactic, butyric, succinic acids – see Equation 2.2), ketones (glycerol, acetone) and alcohols (ethanol, methanol – see Equation 2.3) (Bharathiraja et al., 2018; Tantikhajorngosol et al., 2019).

$$C_6H_{12}O_6 + 2H_2 \rightarrow 2CH_3CH_2COOH + 2H_2O. \qquad (2.2)$$

$$C_6H_{12}O_6 \rightarrow 2CH_3CH_2OH + 2CO_2. \qquad (2.3)$$

2.4.3 Acetogenesis

Acetogenic bacteria attack acidogenesis intermediates, producing acetic acid, CO_2 and H_2. Equations (2.4)–(2.7) depict the reactions that take place during acetogenesis:

$$CH_3CH_2COO^- + 3H_2O \rightarrow CH_3COO^- + H^+ + HCO_3 + 3H_2, \qquad (2.4)$$

$$C_6H_{12}O_6 + 2H_2O \rightarrow 2CH_3COOH + CO_2 + 4H_2, \qquad (2.5)$$

$$CH_3CH_2OH + 2H_2O \rightarrow CH_3COO\text{-} + 2H_2 + H^+, \qquad (2.6)$$

$$2HCO_3^- + 4H_2 + H^+ \rightarrow CH_3COO^- + 4H_2O. \qquad (2.7)$$

Syntrophobacter wolinii, propionate decomposer *Syntrophomonas wolfei*, butyrate decomposer *Clostridium* spp., *Peptococcus* anaerobes, *Lactobacillus* and *Actinomyces* are bacteria that contribute to acetogenesis (Bharathiraja et al., 2018; Abdul Aziz et al., 2019; Abdelsalam et al., 2019).

2.4.4 Methanogenesis

The methanogenesis phase is the final stage of anaerobic digestion. The intermediate products from the other phases are used, with methane being the primary result. The reactions are shown in Equations (2.8)–(2.13). Methanogenesis is caused by a variety of bacteria, including *Methanobacterium*, *Methanobacillus*, *Methanococcus* and *Methanosarcina*, and these are all the bacteria that produce methane.

$$2CH_3CH_2OH + CO_2 \rightarrow 2CH_3COOH + H_2, \qquad (2.8)$$

$$CH_3COOH \rightarrow CH_4 + CO_2, \qquad (2.9)$$

$$CH_3OH \rightarrow CH_4 + H_2O, \qquad (2.10)$$

$$CO_2 + 4H_2 \rightarrow CH_4 + 2H_2O, \qquad (2.11)$$

$$CH_3COO^- + SO_4^{2-} + H^+ \rightarrow 2HCO_3^- + H_2S, \qquad (2.12)$$

$$CH_3COO^- + NO^- + H_2O + H^+ \rightarrow 2HCO_3 + NH_4^+. \qquad (2.13)$$

As you can see, anaerobic digestion bacteria are distinct from other enzymes used to produce biofuels, and they may even be found in our own stomachs! Manure and litter, food wastes, green wastes, plant biomass and wastewater sludge are all examples of organic matter that can be fed to an anaerobic digester (Bharathiraja et al., 2018; Chowdhury et al., 2020; Tamburini et al., 2020). Polysaccharides, proteins and fats/oils are among the elements that make up these feedstocks.

2.5 VARIABLES INFLUENCING ANAEROBIC DIGESTION

Anaerobic digestion may convert almost all organic matter (OM) to biogas; however, some materials produce more methane than others. This brief explains how three metrics such as volatile solids (VS), oxygen demand (OD) and biochemical methane potential (BMP) are used to reduce the spectrum of methane-producing materials from potential to feasible or probable. However, before we begin, we must first clarify a few concepts. Manure is the primary digestion substrate if the digester is utilized to treat manure first. A co-digestion substrate or co-digestion product is an extra organic matter added to a manure digester to improve gas production (Bharathiraja et al., 2018).

2.5.1 VOLATILE SOLIDS CONTENT

The total amount of organic matter in a substrate is determined by using volatile solids analysis. It is a straightforward, low-cost method that provides a mass-based assessment of organic matter. It is probably better to look for another candidate for digestion unless a substrate contains more than 60% or 70% VS on a dry mass basis (Hamilton, 1972).

2.5.2 OXYGEN DEMAND

Organic matter is not all made equal. High-energy organic matter creates more methane than low-energy organic matter. The energy content of organic materials is estimated using the oxygen demand (OD).

Let's go through the definition of oxygen demand once more. The quantity of oxygen required to decompose a pollutant aerobically is used to estimate its organic matter concentration (Hamilton, 1972).

$$OM + O_2 \rightarrow CO_2 + H_2O + energy. \tag{2.14}$$

The oxygen demand of the organic matter (OM) is the mass of O_2 required to convert OM to CO_2 and H_2O. Equation (2.14) also indicates that the amount of energy released during aerobic digestion is proportional to the amount of oxygen consumed.

$$OM + heat \rightarrow CH_4 + CO_2 + H_2O + energy. \tag{2.15}$$

However, we may still utilize OD to calculate the OM's energy potential. On the right-hand side of Equation (2.15), OD must equal OD on the left-hand side of the equation. The majority of anaerobic digestion's energy is stored as CH_4. Because methane (CH_4) is flammable, it has a high oxygen demand:

$$CH_4 + 2O_2 \rightarrow CO_2 + 2H_2O + heat. \tag{2.16}$$

The unit mass of each constituent in methane combustion can be calculated using the knowledge of chemistry (Equation 2.16). On both sides of Equation (2.17), unit masses must be balanced by putting the equation into numbers:

$$16g + 2 \times 32g = 44g + 2 \times 18g. \tag{2.17}$$

One mole of methane weighs 16 g. The mass of two moles of O_2 is 64 g. As a result, 1 g of methane equals 4 g of OD (64 g/16 g = 4 g). In the right-hand side of

Equation (2.16), each gram of oxygen demand eliminated from the OM must be matched with 0.25 g of CH_4. This can be taken a step further. 0.25 g of CH_4 takes up 370 ml under typical conditions (1 atmospheric pressure and 20°C). Rounding up, a gram of substrate OD eliminated generates around 400 ml of methane, depending on reactor temperature (Hamilton, 1972). To put it in another way:

$$1\text{kg Oxygen Demand Removed} \cong 0.4\text{m}^3 \text{ methane produced.} \quad (2.18)$$

The ultimate methane yield is defined as the volume of methane produced per mass of oxygen demand eliminated. The yield of methane is not the same as the yield of biogas. Methane, carbon dioxide, water vapour and a few other gases, notably H_2 and H_2S, make up biogas created from manure. The total amount of biogas emitted per kg of OD removed will exceed the 0.40 m^3 of CH_4 released. Because the majority of the energy in biogas is CH_4, you can still use Equation (2.18) to calculate the potential power released during anaerobic digestion (Hamilton, 1972).

It's also worth noting that Equation (2.18) estimates the volume of CH_4 emitted per mass of oxygen demand reduced. The oxygen demand included in OM is eliminated during digestion, and each material's digestibility varies. Nonetheless, evaluating the oxygen consumption of two materials allows us to compare their biogas production potential. For anaerobic digestion, the chemical oxygen demand (COD) is commonly employed to calculate OD. It's a straightforward test that evaluates OM in the absence of oxygen. However, there are a couple of flaws. Even if the compounds in the sample are poisonous or indigestible to microorganisms, a digesting substrate will have a COD (Hamilton, 1972). In addition, the amount of oxidant consumed in the COD test is proportional to digesting time. The oxygen demand will be underestimated if the analyst does not dilute the sample properly or if the test is halted before all of the oxidant has been used. The maximum amount of methane we may expect from OM is determined by oxygen demand.

2.5.3 BIOCHEMICAL METHANE POTENTIAL

The biochemical methane potential (BMP) analysis is a rather straightforward method despite its sophisticated sounding name: anaerobic microbes are seeded with a sample of substrate, which is then mixed with a nutritional solution and cultured for 30–40 days. During the incubation time, the volume of CH_4 generated is measured. There are several methods to interpret BMP test findings. Specific methane yield, or the volume of CH_4 produced per mass of volatile solids (VS) supplied, is the most usual interpretation (Hamilton, 1972). Wet mass methane potential, which is the amount of CH_4 produced per mass of wet or "as is" sample, is another metric. The volume of CH_4 generated divided by the volume of substrate added is the volumetric methane potential (Hamilton, 1972).

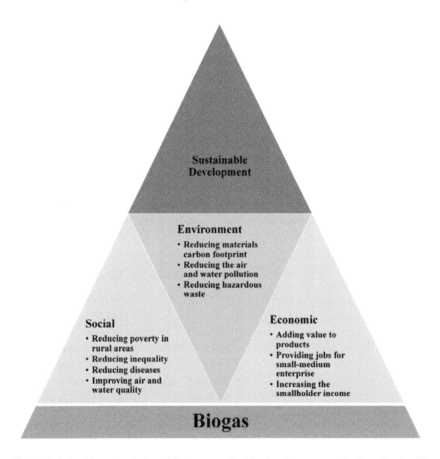

FIGURE 2.1 Biogas' role in achieving sustainable development goals. Reprinted with permission from Obaideen et al. (2022).

2.6 BIOGAS CONTRIBUTION TO SUSTAINABLE DEVELOPMENT GOALS (SDGS)

Biogas contributes to the three pillars of sustainable development goals, namely economic, social and environmental, which is depicted in Figure 2.1. The contribution of biogas to the SDGs and sustainable development dimensions (SDDs) has been discussed in detail in Figure 2.1.

2.6.1 INCREASING RENEWABLE ENERGY PRODUCTION (SDG 7: AFFORDABLE AND CLEAN ENERGY)

Coal, natural gas and petroleum products are the most common sources of heat and power generation in the world. Biogas, which is made from organic waste and agricultural by-products, can be used to replace fossil fuels and reduce CO_2 emissions. Biogas is a sustainable energy source that is not limited by geography and requires sophisticated technologies (Obaideen et al., 2022; Balat & Balat, 2009). Biogas encourages decentralization and democratization of energy generation. Rural and remote villages that aren't connected to power or gas grids can generate their own energy (from biogas) by recycling their garbage. As a result, they will be able to become energy/resource self-sufficient. When there is excess biogas, it can be kept inside a digester, in a small-scale gasholder or pumped into an existing gas infrastructure. As a result, it can be utilized in conjunction with other energy-generating technologies to meet both baseload and peak energy demands. Although biogas has a good role as an affordable clean energy source, it also has drawbacks, such as methane's participation in greenhouse gases, impurities that can cause equipment to degrade or even fail, poor energy density, and volatility in quantity and composition, to name a few (Obaideen et al., 2022).

2.6.2 REDUCING CLIMATE CHANGE IMPACTS (SDG 13: CLIMATE ACTION)

One of the world's most significant environmental challenges is global warming. Unsustainable oil consumption in the past has resulted in global change, which must be addressed (Obaideen et al., 2022; Bilen et al., 2008; Pei-dong et al., 2002). Biogas, unlike fossil fuels, has minimal negative environmental consequences and emits few hazardous emissions. Fossil fuels contribute to greenhouse gas (GHG) emissions, which are the primary cause of climate change, global warming and polar ice melting. Fossil fuels and land-use change each generate 33 × 10^{15} and 38 × 10^{15} tons of GHG per year, respectively (Omer, 2012). By using anaerobic digesters, such as household digesters, desertification, greenhouse gas pollution, soil deterioration and depletion of cultivated land can all be reduced. Biogas is a more environmentally friendly kind of energy because it does not emit any dangerous or damaging gases. Because biogas is a localized energy source, there will be no environmental impact from its transit. Plants will consume this biomass during photosynthesis, and this biomass will be used for biogas generation, as CO_2 is produced during the combustion of biogas. As a result, the CO_2 emission cycle of biogas and other biomass resources is closed or virtually closed. Biogas instead of fossil fuels is predicted to cut GHG emissions by 60–80% in the transportation sector (Obaideen et al., 2022). The use of biogas for local community energy generation and space heating can cut global CO_2 emissions by 18–20%. According to the Intergovernmental Panel on Climate Change (IPCC), the transportation sector accounts for 14% of global GHG emissions (Allen et al., 2018). Livestock emissions, which largely consist of carbon dioxide, methane and nitrous oxide, account for 14.5% of total anthropogenic GHG emissions, according to the FAO data (Gerber et al., 2013). The treatment of manures with anaerobic digestion (AD) reduces the generation of nitrous oxide and absorbs methane as biogas, which can be utilized to generate energy. Hamburg (1989) did a pilot-scale study on the emission of H_2S and SO_2 from cooking with crop stalks, coal and biogas in the Chinese Province of Henan. The SO_2 emissions from crop stalks and coal were four times higher than those from biogas, according to the study's findings. In the case of the biogas, no detectable quantities of H_2S were observed. Yu et al. (2008) examined rural energy generation in China using household-scale biogas digesters and greenhouse gas emission reduction in a systematic study. The pollution from straw, fuel wood, biomass, refined oil, power, liquefied petroleum gas (LPG), natural gas and coal gas is in relation to biogas emissions (Obaideen et al., 2022). In the kitchen, cattle dung is typically used as manure or dung cakes, which are neither sanitary nor cost-effective (Singh & Sooch, 2004). Burning dung cakes pollutes the environment and wastes a valuable fertilizer (if applied directly to the soil) (Bala & Hossain, 1992). The removal of animal faeces by anaerobic digestion is a safe and profitable procedure (Abdelsalam et al., 2016). According to Hiremath et al., 2010, India's energy needs can be met through decentralized energy planning and the use of locally accessible resources. Biogas is one of India's possible energy-generation solutions.

2.6.3 REDUCING AIR AND WATER POLLUTION (POLLUTION PREVENTION) (SDG 3: GOOD HEALTH AND WELL-BEING, SDG 14: LIFE BELOW WATER, SDG 15: LIFE ON LAND)

The biogas industry can help to minimize air and water pollution, which is mostly due to local biogas/biomethane production, eliminating the need for shipping, which has a negative influence on aquatic life (Obaideen et al., 2022; Garg et al., 2005). Biogas can be used to substitute fossil fuels in gas power plants, carry fuel in gas vehicles and collect pollutants from municipal solid waste (MSW) and convert it into valuable energy. The combustion of fossil fuels releases a vast amount of fine particulate matter into the atmosphere, thereby polluting the air. Replacement of fossil fuels, particularly coal, might reduce CO_2 emissions and fine particulate matter greatly. Reduced air pollution also implies less acid rain, which means less water pollution. Oxygen-demanding bacteria in wastewater can lower the oxygen content in surface water. The production of biogas from wastewater not only minimizes wastewater pollution, but the slurry collected from the biodigester can also be used as fertilizer due to its high nitrogen and phosphorus content (Obaideen et al., 2022; Linderson et al., 2002; Weiland, 2010).

Reduced pollution, digested recirculation of nutrients and the use of biogas for power generation would result from the extraction of landfill gas from active and closed landfills and the conversion of excess organic waste to anaerobic digestion (AD) (Obaideen et al., 2022; Mambeli et al., 2014; Zhang & Xu, 2020; Bremond et al., 2021). Furthermore, firewood, crop residues, dried animal dung and crop waste as home fuel are the main sources of indoor air pollution.

Furthermore, firewood is one of the primary causes of deforestation, which results in the accumulation of greenhouse gases. Biogas produced externally from the digestion of home and agricultural waste can be used as a cooking fuel to reduce interior air pollution and deforestation (Semple et al., 2014). The conversion of the organic component of industrial effluents, such as those from palm oil mills, breweries, slaughterhouses and so on, into biogas reduces the environmental impact of these processes while producing energy for their operations, increasing their sustainability and self-reliance (Hosseini & Abdul Wahid, 2015; Tan & Lim, 2019). AD of biosolids promotes a sanitary and hygienic environment by offering decentralized and local care for these wastes. This aids in the prevention of diar-rhoeal disorders caused by bacterial infections, such as cholera, trachoma, schis-tosomiasis and hepatitis (Zinatizadeh et al., 2012; Hosseini & Wahid, 2013). One of the most significant challenges with biogas is the presence of pollutants such as sulphur compounds, siloxanes, halogens and so on. Even after biogas purification, the presence of these contaminants will cause corrosion of the engine and other metallic elements. Maintenance and replacement of damaged parts eventually resulted in additional costs (Khan et al., 2012; Qian et al., 2012). To remove such pollutants conveniently, cheaply and effectively, novel pretreatment procedures are necessary.

2.6.4 IMPROVE AGRICULTURE PRODUCTIVITY AND REDUCE LAND-USE CHANGE (SDG 2: ZERO HUNGER AND SDG 15: LIFE ON LAND)

Bioenergy has enormous potential, but it is not unlimited in the same way that biomass production is. The accessible acreage is not limitless, but rather limited. Most governments have set political targets for increasing the share of renew-able energy in the future. This competition has exacerbated the issue of land use (Obaideen et al., 2022; Knapp et al., 2010). Land use is critical because it helps control and sustain numerous natural processes such as carbon fixation, water maintenance, nutrient cycle, hazardous matter decomposition and genetic resource conservation (Knapp et al., 2010; Bai et al., 2008). Chemical fertilizers are harmful in terms of disposal and land fertility since they destroy pests and soil-friendly bacteria. Because land is already scarce, soil infertility can lead to a food crisis. Soil degradation has recently had a negative impact on territorial ecosystems and agriculture crop yields. To meet the energy demand for cooking, electrification and transportation, mining is done to extract gas and fossil fuels (Obaideen et al., 2022; Barbir et al., 1990).

Also, large-scale biodiesel and bioethanol production based on energy crops and vegetable oil results in land-use conflicts and food scarcity. Using first-gen-eration feedstocks for bioethanol and biodiesel, such as sugarcane and seed oils, could result in food scarcity. Crop residue such as wheat straws, rice husks and cotton sticks add nutrients and organic carbon to the soil when left in the field. When there is no residue left in the field and it is used for bioenergy produc-tion, top fertile soil may be eroded. Reduced soil fertility will decrease aver-age food crop yields, perhaps leading to a shortage of cereals, grains and food (Sarkar et al., 2020). All of these land-related difficulties can be solved with bio-gas generation. There is no need for land mining to extract fossil fuels because biogas may be used as cooking gas, transportation fuel and electricity generation. Biogas derived from animal dung can be used as a transportation fuel in place of bioethanol and biodiesel, both of which are problematic owing to land conflict with food crops. As a by-product of AD, the slurry is utilized as a biofertilizer, increasing soil fertility and agricultural production (Yu et al., 2010; Garg et al., 2005; Igli-nski et al., 2012). Instead of energy-intensive chemical fertilizers, the solid and liquid fractions left over after digesting the feedstock can be utilized on farms as an organic modification/soil improver (digestate or composted digestate biofertilizer). The use of digestate biofertilizer has been shown to boost crop yields (Lesueur et al., 2016; Raimi et al., 2012; Kok et al., 2018; Yong et al., 2021; Obaideen et al., 2022).

2.6.5 ENHANCING THE WASTE MANAGEMENT PROCESS (SDG 11: SUSTAINABLE CITIES AND COMMUNITIES, SDG 12: RESPONSIBLE CONSUMPTION AND PRODUCTION)

Domestic solid biomass fuel-burning pollution has caused approximately 4 million mortalities and 110 million life-year-adjusted impairments. The associated emissions include black carbon, a short-lived climatic pollutant known to disrupt monsoons and hasten glacier melting, endangering people's access to water and food. Residents' exposure to a number of these toxins will be reduced by burning biogas instead of biomass, hence improving their well-being and health (Obaideen et al., 2022; Bonjour et al., 2002; Pujara et al., 2019). Advances in industrial processes, increased commercialization and urbanization resulted in a diverse chemical and solid waste stream, which has proven difficult to dispose safely (Knickmeyer, 2020). Municipal solid waste (MSW) sustainability is defined as the collection of MSW and its proper disposal or conversion into valuable goods (Phu et al., 2020; Zhang et al., 2021). The large volume of waste allows for direct land application, construction material or anaerobic digestion to produce biogas (Pandyaswargo et al., 2019). Waste-to-energy conversion schemes can be aided by government incentives. Rather than dumping garbage away, people will sell it to waste-to-energy treatment plants (Pandyaswargo et al., 2019). The hygienic environment and protection from bacterial infection are improved by decentralized and local waste treatment to produce biogas through anaerobic digester AD. Many dangerous infections, such as dengue fever, can be spread through improper garbage management.

2.6.6 CREATING JOBS, IMPROVING ECONOMIC DEVELOPMENT AND ADDING VALUE TO PRODUCTS (SDG 9: INDUSTRY, INNOVATION AND INFRASTRUCTURE, AND SDG 8: DECENT WORK AND ECONOMIC GROWTH)

Biogas offers a variety of consequences that can help boost economic growth, as seen in Figure 2.2. Solid biomass fuels, such as dried dung cakes, firewood, crop residues, heat and cooking straw, or other agricultural residues, are projected to be used by approximately 3 billion people globally (Obaideen et al., 2022; Caubel et al., 2020). Women and children in underdeveloped nations bear the burden of

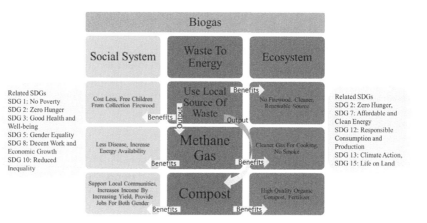

FIGURE 2.2 Impacts of biogas production on the social system and ecosystem with the related SDGs. Reprinted with permission from Obaideen et al. (2022).

obtaining firewood and being exposed to air pollution from domestic cooking. The adoption of AD and biogas as a residential fuel could improve the quality of life in rural areas, and it is already widely used in places like Bangladesh and India. Biogas is an important source of energy for economic growth (Korberg et al., 2020). By adding value to waste, it can be transformed from a government burden into an opportunity to produce biogas, biofertilizer and new jobs. Biogas, for example, has created around 335,000 temporary construction employment and 23,000 full-time operational positions, although it represents less than 2.2% of total renewable energy in 2019 (according to the American Biogas Council (ABC)). (IRENA, 2019). Biogas alone employs 209,000 people in China (IRENA, 2019), and it has a lot to develop in the next few years.

2.6.7   TREATING WASTEWATER (SDG 6: CLEAN WATER AND SANITATION)

Purification of seawater (Obaideen et al., 2022; Sayed et al., 2021a; Al Radi et al., 2021), reverse osmosis process (Fonseca et al., 2021; Martin-Gorriz et al., 2021), desalination systems (Patel et al., 2021; Hamdan et al., 2021), urea fuel cells (Sayed et al., 2021b, 2021c) and bioelectrochemical systems (Sayed et al. 2021d, 2021e) are examples of water purification technologies. Anaerobic digestion of wastewater produces biogas, which can be used to power a diesel engine. This electricity will be utilized to power the water purification process, which will render the water safe to consume. A long-term wastewater treatment strategy was a major challenge. Wastewater treatment will minimize GHG emissions, water pollution and aquatic life hazards, as well as enhance people's economy and quality of life in the areas where these wastewater treatment plants are erected. This plan will meet a wide range of social, environmental and economic goals for long-term growth (Gupta, 2020). AD can be used to stabilize sewage sludge before being applied back to agricultural land as a biofertilizer in places where biosolids are collected and treated in wastewater treatment plants (Spinosa & Doshi, 2021;

Asaoka et al., 2021). AD reduces the carbon content in water, making effluent less hazardous to aquatic bodies and life if discharged (Stazi & Tomei, 2018). The digestion of solid organic waste and wastewater will make these treatment facilities self-sufficient in terms of energy and economy. Where collecting infrastructure is unavailable, small-scale, decentralized care options are being developed and implemented (Obaideen et al., 2022).

2.7   RECENT ISSUES OF BIOGAS PRODUCTION

Recent issues bow down to gap the biotech research and commercialization. Lignocellulose-to-biogas production on a wide scale has a lot of potential, and research into it has previously been done. Two of the key obstacles to AD progress are the difficulties of anaerobic digestion and the risk involved in investing in new technology. The goal of research and development departments in this subject is to improve on anaerobic digestion (AD) technology so that biomethane may be more easily integrated into transportation fuel markets. The criteria to understand the key research biotechnology gap is to analyzed the benefits and cost of production of biogas. Once these criteria have been assessed, they will provide crucial information for determining research priorities for development (Bharathiraja et al., 2018). The type and quantity of microorganisms and/or biocatalysts used for organic waste degradation have an impact on conversion rates and process stability. The cost of biogas production increases as a result of the high manufacturing costs. As a result, current research projects have centred on the development of microorganisms and/or biocatalysts with a broad variety of applications, such as on the improved properties and low production costs of biogas (Weber et al., 2010; Lynd et al., 2005). Electrical power and heat are also required for anaerobic digestion (AD) technology. The most efficient use of utilities is an engineering problem that can be solved in pilot facilities and increase process efficiency. Furthermore, the conversion of lignocellulosic waste into biogas can be integrated with fertilizer production, increasing market competitiveness via by-product (digestate) profits. If advancement in research can make the underdevelopment procedures technologically feasible, then deployment of these technologies will boost conversion efficiency (Blanch, 2012). Microorganism species, pretreatment and purification technologies, substrate properties and optimal reactor conditions are all technical and economic barriers to biogas production. The combination of these steps is key for cost-effective biogas production, and research can help bridge the gap between engineering and biology/technology to develop creative and long-term biogas technology solutions (Blanch, 2012).

2.8   SOME RECENT STUDIES OF BIOGAS PRODUCTION FROM BIOMASS-BASED GREEN CHEMISTRY AS SUSTAINABLE SOURCE FOR ENERGY GENERATION

This section presents work reported on the use of biomass from agricultural waste for biogas production. The availability of fossil fuels is extremely limited, and their combustion is a major source of pollution in the environment. As a result, scientists

TABLE 2.1
The Amount of Biogas Produced from Different Substrates

Substrates	DM [%]	Biogas yield [m³kg⁻¹ of TS]
Excreta from cattle (fresh)	25–30	0.6–0.8
Excreta from sheep (fresh)	18–25	0.3–0.4
Excreta from poultry	10–29	0.3–08
Blood liquid	18	0.3–0.6
Rumen content	12–16	0.3–0.6

Reprinted with permission from Afazeli et al. (2014).

are frantically looking for alternatives to fossil fuels and biomass could be a feasible option. One approach to convert biomass to biogas is by anaerobic digestion. Among the organic waste types used to make biogas are slaughterhouse wastes and animal husbandry residues. Large amounts of livestock waste outputs and slaughterhouse waste materials are produced every year around the world, causing pollution and causing significant concern. In light of what has been said, using animal waste to make biogas and so minimize pollution is a good idea. As a result, researchers are looking into the possibility of producing biogas from animal waste.

TABLE 2.2
Results of Higher CH₄ and CO₂ Yields from Bioreactors

EXP. No	Content (Biomass and RSU)	CH_4 (%)	CO_2 (%)	Time (Days)	Vol. (ml)
1	Potato peel, sediment, and water	35.64	42.22	19	500
2	Papaya peel, sediment, and water	1.64	12.35	10	500
3	Pineapple peel, sediment, and water	0.44	98.22	10	400
4	Pea shell, sediment, and water	16.13	24.57	21	500
5	Banana peel, sediment, and water	0.39	93.7	19	400
6	Bean peel, sediment, and water	0.76	78.53	21	400
7	Sugarcane bagasse, sediment, and water	80.85	17.43	78	625
8	Wet sugarcane bagasse, sediment, and water	96.06	3.57	78	675
9	Semi-dry sugarcane bagasse, sediment, and water	91.39	8.17	78	750
10	Potato peel, sugar bagasse, sediment, and water	58.74	26.46	78	350

Note: The CH_4 and CO_2 yields of the ten experiments were obtained by chromatographic analysis in the established period of 7–78 days. The difference in days was the result of the variation of biogas content generated by each experiment.
Reprinted with permission from Bonilla et al. (2020).

2.8.1 An Investigation of Biogas Production Potential from Livestock and Slaughterhouse Wastes

Afazeli et al. (2014) investigated the slaughterhouse waste, which includes rumen, intestines, stomach and blood from both heavy and light livestock, as well as chicken blood in Iran. According to the findings, the available livestock dung has the capacity to produce 8,600 million m^3 of biogas per year, with 70% of that coming from heavy livestock, 23% from poultry and just 7% from light livestock. This indicates that in Iran, biogas yields potential from slaughterhouse waste is roughly 54 million m^3 per year, with 40% coming from light livestock rumen, 24% from heavy livestock rumen, 17% from heavy livestock blood, 14% from chicken blood and 5% from light livestock blood, and the data are provided in Tables 2.1 and 2.2.

2.8.2 Methane Gas Generation through the Anaerobic Co-digestion of Urban Solid Waste and Biomass

Bonilla et al. (2020) reported methane gas generation by anaerobic co-digestion of biomass and solid urban waste. The authors applied gas chromatography for the determination of CH_4 and CO_2 percentage, which has been displayed in the Tables 2.2 and 2.3 and Figures 2.3 and 2.4. Their findings show that the optimum methane gas produced was on sugarcane bagasse according to tables and figures mentioned earlier.

The calculated energy potential values of the biogas generated from sugarcane bagasse, sugar solid urban waste sediment and water are listed in the table mentioned below using Equation (2.19) with caloric value of biogas ranging from 19.6 to 25 MJ/m^3.

EXP No	Content (Biomass and RSU)	CH$_4$ (%)	CO$_2$ (%)	Time (Days)	Energy Potention (MJ/m³)
8	Wet sugarcane bugasse, sediment, and water	96.06	3.57	78	15.05
9	Semi-dry sugarcane bagasse, sediment, and water	91.39	8.17	78	16.72

$$\text{Energy potential}(MJ) = \frac{m^3 \text{Biogas obtained} \times \text{caloric capacity of biogas}(MJ)}{1\,m^3 \text{biogas}}.$$

(2.19)

TABLE 2.3

Experimental Design in Triplicate of the Experiments with the Highest Performance on a Laboratory Scale in a Digestion Time of 6–78 Days

EXP No	REP No	CH$_4$ (%)	CO$_2$ (%)	Time (Days)	Vol. (ml)	EXP No	REP No	CH$_4$ (%)	CO$_2$ (%)	Time (Days)	Vol. (ml)
8	1	18.89	29.9	6	100	9	1	39.47	54.54	6	200
		80.09	6.44	24	175			47.38	20.2	24	250
		89.7	8.71	48	475			83.07	16.68	48	450
		96.06	3.57	78	675			91.39	8.17	78	750
	2	16.05	28.19	6	80		2	37.75	52.65	6	150
		78.1	4.33	24	145			44.46	17.06	24	225
		85.34	6.78	48	395			79.89	11.59	48	455
		93.79	2.01	78	545			85.89	8.03	78	785
	3	17.9	30.12	6	125		3	29.58	65.01	6	230
		79.02	5.01	24	200			46.79	23.8	24	275
		87.46	7.89	48	450			81.45	17.89	48	480
		95.23	2.2	78	700			89.9	8.69	78	730

Reprinted with permission from Bonilla et al. (2020).

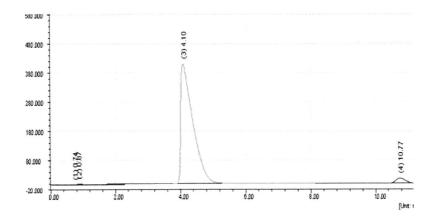

FIGURE 2.3   Chromatogram with higher CH_4 and CO_2 performance from Experiment 8.

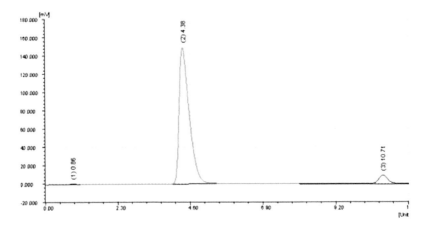

FIGURE 2.4   Chromatogram with higher CH_4 and CO_2 yield from Experiment 9. Reprinted with permission from Bonilla et al. (2020).

2.8.3   IMPROVEMENT OF BIOGAS QUALITY PRODUCT FROM DAIRY COW MANURE USING NaOH AND Ca(OH)₂ ABSORBENTS ON HORIZONTAL TUBE FILTRATION SYSTEM OF MOBILE ANAEROBIC DIGESTER

Herwintono et al. (2020) carried out an experiment on the improvement of biogas quality product from dairy cow manure using NaOH and $Ca(OH)_2$ as absorbents on a horizontal tube filtration system of mobile anaerobic digester, as shown in Figure 2.5. The authors recorded the average temperature inside and outside the digester to be 28.37°C and 22.16°C, respectively. The average biogas temperature before filtration 28.37°C and after filtration 31.09°C was also noted. Based on their findings, the absorbents NaOH and $Ca(OH)_2$ used for the study could increase the content of methane and oxygen gas and reduced carbon (iv) oxide gas. More details of the results are illustrated in Tables 2.4 and 2.5.

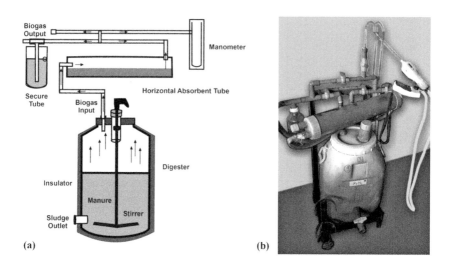

(a)   (b)

FIGURE 2.5 (a) The diagram model of mobile anaerobic digester and (b) a unit of mobile anaerobic digester. Reprinted with permission from Herwintono et al. (2020).

TABLE 2.4
Inside and Outside Temperature of Digester and Biogas Temperature before and after Filtration

Type of Absorbent	Digester temperature (°C)		Biogass temperature (°C)	
	Outside	Inside	Before filtration	After filtration
Control	27.12	28.34	28.79	29.21
NaOH	27.10	28.23	28.09	33.64
Ca(OH)$_2$	27.28	28.40	28.26	31.99
Average	27.16 ± 0.08	28.36 ± 0.07	28.37 ± 0.30	31.61 ± 1.83

Reprinted with permission from Herwintono et al. (2020).

2.8.4 SERIAL ANAEROBIC DIGESTION IMPROVES PROTEIN DEGRADATION AND BIOGAS PRODUCTION FROM MIXED-FOOD WASTE

To ensure effective degradation and high methane yield, as well as to limit methane emissions from the digestate, the biogas generation process must be optimized. The goal of this study was to compare serial digester systems with two or three biogas reactors to a single reactor in order to improve the degree of degradation and methane yield from food waste, as well as assess microbial community adaptation to different reactor steps (Perman et al., 2022). According to the authors' findings, the total organic load (2.4 g VS/(L d)) and hydraulic retention time were the same in all systems (55 days). When compared to a single reactor, serial systems enhanced methane yield by over 5%, with the first-step reactors

TABLE 2.5

Biogas Contents before and after Filtration Using NaOH and Ca(OH)$_2$

Gas contents (%)	NaOH			Ca(OH)$_2$		
	Before filtration	After filtration	Percntage of increasing (+)/decreasing (−)	Before filtration	After filtration	Percntage of increasing (+)/decreasing (−)
CH$_4$	67.48 ± 1.06	77.07 ± 0.87	+14.26	68.62 ± 0.82	73.95 ± 2.27	+7.77
CO$_2$	30.91 ± 1.11	21.37 ± 0.84	−30.86	29.98 ± 0.82	24.62 ± 2.24	−17.88
O$_2$	0.35 ± 0.06	0.42 ± 0.07	+22.06	0.25 ± 0.10	0.27 ± 0.34	+11.22

Reprinted with permission from Herwintono et al. (2020).

TABLE 2.6A

Process Parameters for the Laboratory-Scale Test Reactors Operating in Monodigestion or in Series with Two or Three Reactors

System	Reactor	HRT (days)	OLR (g VS/ (L/d))	Reactor volume (L)	Temperature (°C)	Reactive volume (%)
One-step	A1	55	2.43	9.1	42	100
Two-step	B1	35	3.82	9.1	42	64
	B2	23		4	41	36
Three-step	C1	23	5.74	9.1	42	42
	C2	21		6.3	42	35
	C3	12		3.3	41	23

Reprinted with permission from Perman et al. (2022).

producing the majority of the methane. Serial systems also resulted in improved protein breakdown, with >20% lower outgoing protein concentrations compared to single reactors and increasing NH_4^+–N concentrations with each reactor phase. Parameters for laboratory-scale test reactors that operate in monodigestion or in series with two or three reactors are listed in Table 2.6a.

Analytical information for the digestate and gas produced by various test reactors is shown in Table 2.6b. Reactor A1 was used as a single reactor, whereas reactors B1 and B2 were used in a two-step serial system, and reactors C1, C2 and C3 were used in a three-step serial system. Statistical significance is shown by letters (values with different letters differ significantly (p 0.05)). The columns are compared one by one.

Wannasek et al. (2017) suggested *Sorghum* as a sustainable feedstock for biogas production. The impact of climate variety and harvesting time on maturity and biomass yield was also investigated by Wannasek et al. (2012). Three vegetative phases of five Sorghum types were used in their experiments. Their study provides new information on the development of biomass yield, maturity level and biogas production. When *Sorghum* was cultivated as the principal crop, the dry matter (DM) biomass production ranged from 15.7 to 20.62. Their research presents a correlation between yields and growing degree days (GDD) values over the course of the vegetative period, allowing yield and economic efficiency projections to be transferred from current results to other sites. From their findings when *Sorghum* is utilized as a catch crop it produces excellent yields, a *Sorghum* variety with 1,100 growing degree days (GDD), and has proven to be a fast-ripening variety that is especially excellent as a summer catch crop. Crop maturities of 20–45% DM were studied in this study. Within this range, there was no evidence of a decline in specific methane yield as maturity and DM yield increased. One year with little precipitation was included in the three-year trial data. The statistical study revealed no significant influence of precipitation on biomass yield,

TABLE 2.6B

Analytical Data for the Digestate and Gas Produced from Different Test Reactors

Reactor	NH_4^+–N (mg/kg)*	NH_3–N (mg/kg)*	pH*	VS in digestable (%)*0	VS reduction (%)*	Methane content (%)**	SMP complete system (L CH_4/kg VS)**	SMP front-step reactors (L CH_4/kg VS)**	Methane production (%)***
A1	3.0 (0.1)	287 (9.5)	7.8 (0.0)	3.2 (0.0	78.9 (0.3)	61.8 (0.5)	489 (24)	409	100
B1	2.7 (0.1)	261 (5.6)	7.8 (0.0)	3.4 (0.0)	77.3 (0.3)	65.4 (0.2)	517 (20)	499	96.3
B2	3.4 (0.1)	395 (6.4)	7.9 (0.1)	2.8 (0.1)	81.0 (0.4)	59.9 (1.1)		ND	3.7
C1	2.4 (0.0)	200 (24.8)	7.7 (0.0)	3.8 (0.1)	73.4 (0.4)	62.2 (0.2)	515 (14)	475 (14)	92.3
C2	3.3 (0.1)	384 (6.9)	8.0 (0.1)	3.0 (0.0)	79.1 (0.3)	64.2 (0.5)		ND	6.2
C3	3.5 (0.1)	543 (58.5)	8.0 (0.1)	2.8 (0.1)	80.4 (0.7)	59.3 (1.8)		ND	1.5

Reprinted with permission from Perman et al. (2022).

TABLE 2.7
List of *Sorghum* Crops Applied in the Field Tests

Name	Type	Biomass yield[a]	Maturity tendency[a]	Variety denomination
SOR 1	*Sorghum bicolor*	High	Medium/late	Branco
SOR 2	*Sorghum bicolor × sorghum Sudanese*	Medium	Early/medium	Maja
SOR 3	*Sorghum bicolor*	High	Mediaum/late	Zerberus
SOR 4	*Sorghum bicolor*	High	Late	Bulldozer
SOR 5	*Sorghum bicolor*	High	Late	Sucro sorgho

Reprinted with permission from Wannasek et al. (2017).

confirming *Sorghum*'s drought resistance. According to the authors, the finding is especially significant in light of climate change adaptation. Table 2.7 lists the *Sorghum* species used in the tests. Except for SOR 2, which is a hybrid of *Sorghum bicolor* and *Sorghum sudanese*, other crops are pure *Sorghum bicolor* kinds.

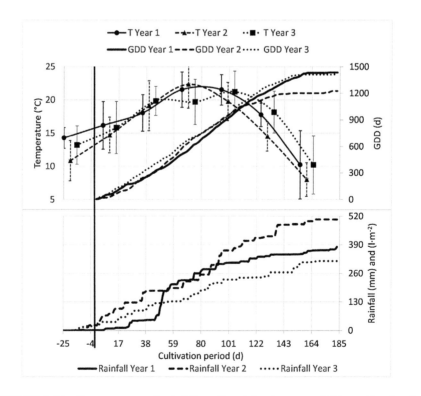

FIGURE 2.6A   Temperature and rainfall profile of the three-year field test. Reprinted with permission from Wannasek et al. (2017).

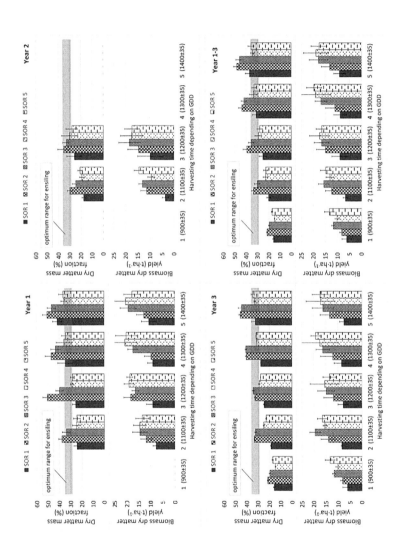

FIGURE 2.6B Maturity, biomass dry matter yield and harvest time of five *Sorghum* cultivars as a function of growing degree days (GDD) and specific dry matter for each cultivation year and the average of the whole period (years 1–3). Reprinted with permission from Wannasek et al. (2017).

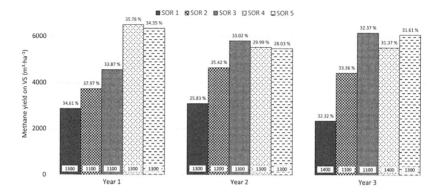

FIGURE 2.6C Comparison of specific methane yield of five *Sorghum* varieties harvested at optimal maturity level. Values above bars are DM%, and values on the bottom are growing degree days (GDD). Reprinted with permission from Wannasek et al. (2017).

Figure 2.6a depicts the cumulative rainfall distribution across the three years of the vegetative period (183 days).

Figure 2.6b shows the maturity and biomass yields as a function of harvesting time for the aforementioned authors.

Figure 2.6c shows the specific methane yield per hectare of five *Sorghum* varieties at optimal maturity (27–35% DM).

Khan and Ahring (2020) researched the effect of pretreatment on anaerobic digestion of recalcitrant fraction of manure for improvement of biogas yield. The effects of various pretreatments on the anaerobic digestion of manure fibres following anaerobic digestion are investigated on the manure fibre fraction separated out after anaerobic digestion of cow manure using physical, chemical and thermal, and thermal combined with alkaline pretreatments using sodium hydroxide were also evaluated. In semi-continuous anaerobic bioreactors, anaerobic digestion after pretreatment was performed alongside untreated controls. All pretreatments had a favourable effect, with the 3% NaOH-pretreated sample at 180°C showing the greatest improvement in VS conversion (42.4%) and methane output (approximately 127%). During pretreatment, cellulose, hemicellulose and lignin in digested manure fibres were reduced by 24.8, 29.1 and 9.5%, respectively, while 76.5% of cellulose and 84.9% of hemicellulose were converted to methane during AD, according to composition analysis Khan and Ahring (2020). Figures 2.7a and 2.7b depict the reduction of cellulose, hemicellulose and lignin during various pretreatments. The sample pretreated at 180°C with the highest dose of NaOH (3%) before digestion had the maximum decrease of cellulose, hemicellulose and lignin, respectively, of 24.8, 29.1 and 9.5%. However, the cellulose reduction is higher, which is most likely owing to the fact that the pretreatment was done at a higher temperature, which has previously been shown in Figure 2.7a.

Pretreatment is required for the conversion of cellulose to biofuel since it is a crystalline polymer protected by lignin. The crystallinity of lignocellulosic materials had previously been discovered to be altered by alkali pretreatment. The overall result of NaOH thermal pretreatment was not only the dissolution of

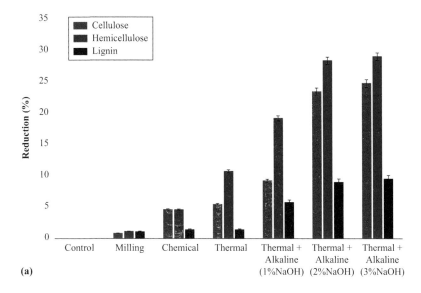

FIGURE 2.7A   Reduction of cellulose, hemicellulose and lignin as the result of different pretreatment conditions. From Khan and Ahring (2020).

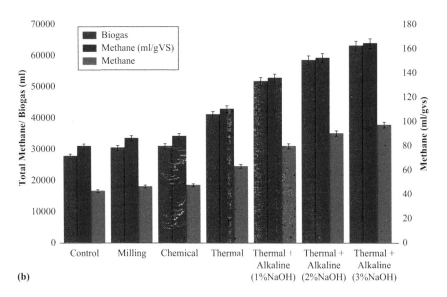

FIGURE 2.7B   Total biogas, total methane and methane yield (ml/gVS) from all the samples. From Khan and Ahring (2020).

portions of the lignin but also the exposure of the carbohydrates fraction, which allowed for substantial biodegradation of this fraction. In line with this, the cellulose and hemicellulose content of all effluent samples dropped (Table 2.8). Figure 2.7b shows the total biogas, total methane and average methane yield (ml/gvs)

TABLE 2.8
Compositional Analysis of Feed and Effluent.

Precondition treatment	Sample type	Cellulose	Hemicellulose	Lignin
Control	Feed	37.1 ± 0.5	26.1 ± 0.4	36.7 ± 0.9
	Effluent	13.8 ± 0.2	9.08 ± 0.1	42.1 ± 0.7
Milling	Feed	36.8 ± 0.3	25.8 ± 0.7	36.3 ± 0.4
	Effluent	12.5 ± 0.7	5.7 ± 0.2	39.6 ± 1.2
Chemical (2%NaOH)	Feed	35.4 ± 1.1	24.9 ± 0.6	36.2 ± 0.4
	Effluent	12.1 ± 0.9	5.4 ± 0.3	38.9 ± 0.6
Thermal	Feed	35.1 ± 0.8	23.3 ± 1	36.2 ± 0.9
	Effluent	12.9 ± 0.3	5.1 ± 0.2	40.8 ± 0.3
Themochemical (1%NaOH)	Feed	33.7 ± 0.6	21.1 ± 1.6	34.6 ± 1.4
	Effluent	7.2 ± 0.4	3.7 ± 0.2	36.3 ± 0.7
Thermochemical (2%NaOH)	Feed	28.4 ± 0.7	18.7 ± 0.9	33.4 ± 0.8
	Effluent	6.7 ± 0.2	3.1 ± 0.1	36.2 ± 0.4
Thermochemical (3%NaOH)	Feed	27.09 ± 0.7	18.5 ± 1.3	33.2 ± 1.1
	Effluent	6.2 ± 0.2	2.6 ± 0.1	34.2 ± 1.3

Reprinted with permission from Khan and Ahring (2020).

data. The findings revealed that all pretreatments improve the anaerobic digestion of digested manure fibres.

2.9   THE KINETICS OF BIOGAS PRODUCTION RATE FROM CATTLE MANURE IN BATCH MODE

Budiyono et al. (2010) studied the kinetic of biogas production using rumen fluid of animal ruminants as inoculums. They inoculated rumen fluid with cattle manure as a substrate into the anaerobic biodigester. The total volume of biogas produced was used to determine the performance of biodigester in their study. The results of their research show that rumen fluid inoculated to biodigester gave a significant effect of two to three times the biogas production compared to manure substrate without rumen fluid. In the presence of rumen fluid, the kinetic parameters of the produced biogas are as follows: the rate constant for the produced biogas (U = 3.89 ml/gvs.day), the maximum biogas produced (A = 172.51 ml/gvs) and the minimum time for the production of biogas (t = 2.25 days). For the substrate without rumen fluid, its kinetic parameters include (U = 1.74 ml/gvs.day), (A = 73.81 ml/gvs) and (t = 14.75 days). The model developed for the study of biogas production in batch mode was given as in Equation (2.19).

TABLE 2.9

The Role of Anaerobic Digestion in the Enhancement of Bioenergy

Material	Contribution	Proposed Model	Reference
Mixed-food waste	Biogas, protein degradation	Anaerobic co-digestion	Perman et al. (2022)
Animal waste, agricultural and household waste	Biogas and electricity	Anaerobic digestion	Singh et al. (2020)
De-oiled rice bran and cattle dung	Biogas and methane	Anaerobic digestion	Jha et al. (2020)
Macro- and micro-algal biomass	Biohydrogen and biomethane	Anaerobic co-digestion	Ding et al. (2016)
Carbohydrate waste feedstock and microbial – *Bacillus firmus*	Biohydrogen	Anaerobic co-digestion	Sinha and Pandy, 2013
Microalgae and carbon-rich co-substrates barley straw	Biomethane	Anaerobic co-digestion	Herrmann et al. (2016)
Sugarcane bagasse	Biohydrogen	Anaerobic co-digestion	Cheng and Zhu (2013)
Sorghum	Biogas	Anaerobic digestion	Wannasek et al. (2017)
Mixture of poultry dropping and *Carica papaya* peels	Biogas	Anaerobic co-digestion, RSM and ANN	Dahunisi et al. (2016)
Cattle manure	Methane	Anaerobic digestion Pilot scale	Tsapekos et al. (2016)
Animal manure	Biogas	Anaerobic digestion	Usman Khan and Kiaer Ahring (2020)
Mixture of sugarcane and rice husk	Biogas	Anaerobic digestion	Usman and Ekwenchi (2013)
Bioplastics	Biogas	Anaerobic digestion	Shrestha et al. (2020)
Agricultural waste	Biogas	Anaerobic digestion	Hernandez and Jimene -2018
Solid urban waste like potato, papaya, pineapple and banana peels	Biogas	Anaerobic digestion	Bonilla et al. (2020)
Food waste like taro, papaya and sweet potato	Methane	Anaerobic co-digestion	Ge et al. (2014)
Livestock manure	Biogas	Anacrobic co-digestion	Afazeli et al. (2014)
Mixture of cladodes of *Opuntia ficus-indica* and cow dung	Biogas and methane	Batch digestion	Jigar et al. (2011)
Variety of cow dung	Biogas	Anaerobic digestion	Godi et al. (2013)
Kitchen waste	Biohydrogen and biomethane	Batch digestion	Sighal et al. 2012

(*Continued*)

TABLE 2.9 (CONTINUED)
The Role of Anaerobic Digestion in the Enhancement of Bioenergy

Material	Contribution	Proposed Model	Reference
Mixture of biomass like bamboo dust, sugarcane bagasse, sawdust and rice husk with cattle dung	Biogas	Batch digestion	Ghatak and Mahanta (2014)
Cow dung, cowpea and cassava peeling	Biogas	Anaerobic digestion	Ukpai and Nnabuchi 2012
Winery waste	Biogas	Mathematical modelling CSTR model	Colussi et al. (2012)
Cow dung and rice husk	Biogas	Anaerobic co-digestion	Iyaba et al. (2009)
Abattoir waste	Biogas	Anaerobic co-digestion	Rabah et al. (2010)
Fruits and vegetable waste with cow dung	Biogas	Anaerobic digestion	Nyong et al. (2021)

$$P = A \exp\left[-\exp\left(\frac{Ue(\lambda - t) + 1}{A} \right) \right] \tag{2.20}$$

where P is the cumulative of specific biogas produced in ml/gvs, A is the production of biogas potential in ml, U is the maximum rate of biogas production ml/gvs .day and λ is the maximum time to produce biogas per day. Nonlinear regression was used in the determination of these parameters according to Budiyono et al. (2010).

Table 2.8 shows the product associated with the role of anaerobic digestion in the enhancement of bioenergy according to the scientific literature (Table 2.9).

2.10   CONCLUSIONS AND PERSPECTIVES

One of the promising renewable energy sources that can significantly lessen the negative environmental effects of fossil fuels is biogas. The raw materials used to produce the biogas have a direct impact on its output, contaminants and composition. Compared to natural gas, biogas provides a number of advantages. As a result, it is today utilized in a wide range of applications, from small-scale use in homes to large-scale use in a number of industries, including power plants. The perspectives to ascertain the role of biogas in achieving the sustainable development goals (SDGs) are as follows: the main contribution comes from increasing the production of renewable energy, decreasing the effects of climate change, preventing pollution and improving agricultural productivity, and reducing land-use change. It also comes from improving waste management, creating jobs, boosting the economy, adding value to products and treating wastewater.

REFERENCES

Abdelsalam, E., Samer, M., Attia, Y. A., Abdel-Hadi, M. A., Hassan, H. E., & Badr, Y. (2016). Comparison of nanoparticles effects on biogas and methane production from anaerobic digestion of cattle dung slurry. *Renewable Energy*, *87*, 592–598.

Abdul Aziz, N. I. H., Hanafiah, M. M., & Mohamed Ali, M. Y. (2019). Sustainable biogas production from agrowaste and effluents-A promising step for small-scale industry income. Renew Energy, *132*, 363–369.

Abdelkareem, M. A., Elsaid, K., Wilberforce, T., Kamil, M., Sayed, E. T., & Olabi, A. (2021). Environmental aspects of fuel cells: A review. *Science of the Total Environment*, *752*, 141803.

Abdelkareem, M. A., Lootah, M. A., Sayed, E. T., Wilberforce, T., Alawadhi, H., Yousef, B. A. A., & Olabi, A. G. (2021). Fuel cells for carbon capture applications. *Science of the Total Environment*, *769*, 144243.

Abdelkareem, M. A., Tanveer, W. H., Sayed, E. T., Assad, M. E. H., Allagui, A., & Cha, S. (2019). On the technical challenges affecting the performance of direct internal reforming biogas solid oxide fuel cells. *Renewable and Sustainable Energy Reviews*, *101*, 361–375.

Afazeli, H., Jafari, A., Rafiee, S., & Nosrati, M. (2014). An investigation of biogas production potential from livestock and slaughterhouse wastes. *Renewable and Sustainable Energy Reviews*, *34*, 380–386.

Akkarawatkhoosith, N., Kaewchada, A., & Jaree, A. (2019). High-throughput CO_2 capture for biogas purification using monoethanolamine in a microtube contactor. *Journal of the Taiwan Institute of Chemical Engineers*, *98*, 113–123.

Alayi, R., Shamel, A., Kasaeian, A., Harasii, H., & Topchlar, M. A. (2016). The role of biogas to sustainable development (aspects environmental, security and economic). *Journal of Chemical and Pharmaceutical Research*, 8(4), 112–118.

Allen, M., Babiker, M., Chen, Y., de Coninck, H., Connors, S., van Diemen, R., & Zickfeld, K. (2018). *Global warming of 1.5 C: Special report on the impacts of global warming*. Geneva: Intergovernmental Panel on Climate Change (IPCC).

Al Radi, M., Sayed, E. T., Alawadhi, H., & Abdelkareem, M. A. (2021). Progress in energy recovery and graphene usage in capacitive deionization. *Critical Reviews in Environment Science and Technology*, *100*, 1–52.

Asaoka, S., Yoshida, G., Ihara, I., Umehara, A., & Yoneyama, H. (2021). Terrestrial anaerobic digestate composite for fertilization of oligotrophic coastal seas. *Journal of Environment Management*, *293*, 112944.

Bala, B., & Hossain, M. (1992). Economics of biogas digesters in Bangladesh. *Energy*, *17*(10), 939–944.

Balat, M., & Balat, H. (2009). Biogas as a renewable energy source – A review. *Energy Sources, Part A: Recovery, Utilization, and Environmental Effects*, *31*(14), 1280–1293.

Bai, Z., Dent, D. L., Olsson, L., & Schaepman, M. E. (2008). Global assessment of land degradation and improvement: 1. Identification by remote sensing. In ISRIC – World Soil Information 2008. https://www.isric.org/sites/default/files/isric- report -2008-01.pdf.

Barbir, F., Veziroğlu, T. N., & Plass Jr, H. J. (1990). Environmental damage due to fossil fuels use. *International Journal of Hydrogen Energy*, *15*(10), 739–749.

Bharathiraja, B., Sudharsana, T., Jayamuthunagai, J., Praveenkumar, R., & Sivasankaran, C., Iyyappan, J. (2018). Biogas production – A review on composition, fuel properties, feed stock and principles of anaerobic digestion. *Renewable and Sustainable Energy Reviews*, *90*, 570–582.

Bilen, K., Ozyurt, O., Bakırcı, K., Karslı, S., Erdogan, S., Yılmaz, M., & Comaklı, O. (2008). Energy production, consumption, and environmental pollution for sustainable development: A case study in Turkey. *Renewable and Sustainable Energy Reviews*, *12*(6), 1529–1561.

Blanch, H. W. (2012). Bioprocessing for biofuels. *Current Opinion in Biotechnology*, *23*(3), 390–395.

Bonilla, H. R., Vega, C., Feijoo, V., Villacreses, D., Pesantez, F., & Olivera, L. (2020). Methane gas generation through the anaerobian codigestion of urban solid waste and biomass. *Energy Report*, *6*, 430–436.

Bonjour, S., Pruss-Ustun, A., & Rehfuess, E. (2002). *Indoor air pollution: National burden of disease estimates.* Geneva: World Health Organization. Retrieved from https://www.who.int/airpollution/publications/indoor.

Bremond, U., Bertrandias, A., de Buyer, R., Latrille, E., Jimenez, J., Escudie, R., … Carrere, H. (2021). Recirculation of solid digestate to enhance energy efficiency of biogas plants: Strategies, conditions and impacts. *Energy Conversion and Management*, *231*, 113759.

Bremond, U., Bertrandias, A., Steyer, J. P., Bernet, N., & Carrere, H. (2021). A vision of European Biogas sector development towards 2030: Trends and challenges. *Journal of Cleaner Production*, *287*, 125065.

Budiyono, I. N., Widiasa, S., & Johari, S. (2010). Kinetic of biogas from cattle manure in batch mode. *International Journal of Chemical and Biological Engineering*, *3*(1), 39–44.

Calbry-Muzyka, A. S., Gantenbein, A., Schneebeli, J., Frei, A., Knorpp, A. J., Schildhauer, T. J., & Biollaz, S. M. (2019). Deep removal of sulfur and trace organic compounds from biogas to protect a catalytic methanation reactor. *Chemical Engineering Journal*, *360*, 577–590.

Caubel, J. J., Rapp, V. H., Chen, S. S., & Gadgil, A. J. (2020). Practical design considerations for secondary air injection in wood-burning cookstoves: An experimental study. *Development Engineering*, *5*, 100049.

Chowdhury, T., Chowdhury, H., Hossain, N., Ahmed, A., Hossen, M. S., Chowdhury, P., … Saidur, R. (2020). Latest advancements on livestock waste management and biogas production: Bangladesh's perspective. *Journal of Cleaner Production*, *272*, 122818.

Cheng, J., & Zhu, M. (2013). A novel anaerobic co-culture system for bio-hydrogen production from sugarcane bagasse. *Bioresource Technology*, *144*, 623–631.

Colussi, I., Cortesi, A., Gallo, V., Rubesa Fernandez, A. S., & Vitanza, R. (2012). Modelling of an anaerobic process producing biogas from winery waste. *Chemical Engineering Transactions*, *27*, 301–306.

Ding, L., Cheng, J., Xia, A., Jacob, A., Voelklein, M., & Murphy, J. D. (2016). Co-generation of biohydrogen and biomethane through two-stage batch co-fermentation of macro- and micro-algal biomass. *Bioresource Technology*. http://doi.org/10.1016/j.biortech.2016.06.092

Dahunsi, S. O., Oranusi, S., Owolabi, J. B., & Efeovbokhan, V. E. (2016). Mesophilic anaerobic codigestion of poultry dropping and *Carica papaya* peels: Modelling and process parameter optimization study. *Bioresource Technology*. http://doi.org/10.1016/j.biortech.2016.05.118

Fonseca, M. J. D. C., Silva, J. R. P. D., Borges, C. P., & Fonseca, F. V. D. (2021). Ethinylestradiol removal of membrane bioreactor effluent by reverse osmosis and UV/H2O2: A technical and economic assessment. *Journal of Environment Management*, *282*, 111948.

Garg, R. N., Pathak, H., Das, D., & Tomar, R. (2005). Use of flyash and biogas slurry for improving wheat yield and physical properties of soil. *Environmental Monitoring and Assessment, 107*(1–3), 1–9.

Gerber, P. J., Steinfeld, H., Henderson, B., Mottet, A., Opio, C., Dijkman, J., ... Tempio, G. (2013). *Tackling climate change through livestock: A global assessment of emissions and mitigation opportunities.* Rome: food and agriculture organization of the United Nations (FAO). Retrieved from http://www.fao.org/.../i3437e00.htm.

Ge, X., Matsumoto, T., Keith, L., & Yebo, L. (2014). Biogas energy production from tropical biomass wastes by anaerobic digestion. *Bioresource Technology, 169*, 38–44.

Ghatak, M. D., & Mahanta, P. (2014). Kinetic assessment of biogas production from lignocellulosic biomasses. *International Journal of Engineering and Advanced Technology, 3*(5), 244–249.

Godi, N. Y., Zhengwuvi, L. B., Adulkadir, S., & Kamtu, P. (2013). Effect of cow dung variety on biogas production. *Journal of Mechanical Engineering Research, 5*(1), 1–4.

Gupta, A. S. (2020). Feasibility study for production of biogas from wastewater and sewage sludge: Development of a sustainability assessment framework and its application. Retrieved June 2020, from http://urn.kb.se/resolve?urn=urn:nbn:se:kth:diva-274370.

Hamburg, R. (1989). Household cooking fuel hydrogen sulfide and sulfur dioxide emissions from stalks, coal and biogas. *Biomass, 19*(3), 233–245.

Hamdan, H., Saidy, M., Alameddine, I., & Al-Hindi, M. (2021). The feasibility of solar-powered small-scale brackish water desalination units in a coastal aquifer prone to saltwater intrusion: A comparison between electrodialysis reversal and reverse osmosis. *Journal of Environment Management, 290*, 112604.

Hamilton, D. W. (1972). Anaerobic digestion of animal manures: Methane production of waste materials. BAE 1760. Division of Agricultural Sciences and Natural Resources. OK. Oklahoma Cooperative extension service.

Hernandez, S. C., & Jimenez, L. D. (2018). The potential for biogas production from agriculture wastes in Mexico. *Intech Open.* http://doi.org/10.5772/intechopen.75457

Herrmann, C., Kalita, N., Wall, D., Xia, A., & Murphy, J. D. (2016). Optimised biogas production from microalgae through co-digestion with carbon-rich co-substrates. *Bioresource Technology, 214*, 328–332.

Herwintono, Winaya, A., Khotimah, K., Hidayati, A. (2020). Improvement of biogas quality product from dairy cow manure usingNaOH and Ca(OH)2 absorbents on horizontal tube filtration system ofmobile anaerobic digester. *Energy report, 6*, 319–324.

Hiremath, R. B., Kumar, B., Balachandra, P., & Ravindranath, N. (2010). Sustainable bioenergy production strategies for rural India. *Mitigation and Adaptation Strategies for Global Change, 15*(6), 571–590.

Hosseini, S. E., & Abdul Wahid, M. (2015). Pollutant in palm oil production process. *Journal of the Air and Waste Management Association, 65*(7), 773–781.

Hosseini, S. E., & Wahid, M. A. (2013). Feasibility study of biogas production and utilization as a source of renewable energy in Malaysia. *Renewable and Sustainable Energy Reviews, 19*, 454–462.

Höfer, R., & Bigorra, J. (2008). Biomass-based green chemistry: Sustainable solutions for modern economies. *Green Chemistry Letters and Reviews, 1*(2), 79–92.

Igliński, B., Buczkowski, R., Iglińska, A., Cichosz, M., Piechota, G., & Kujawski, W. (2012). Agricultural biogas plants in Poland: Investment process, economical and environmental aspects, biogas potential. *Renewable and Sustainable Energy Reviews, 16*(7), 4890–4900.

Iyaba, E. T., Mangibo, I. A., & Mohammad, Y. S. (2009). The study of cow dung as co-substrate with rice husk in biogas production. *Scientific Research and Essays*, *4*(9), 861–866.

Jha, B., Chandra, R., Vijay, V. K., Subbarao, M. V., & Isha, A. (2020). Utilization of de-oiled rice bran as a feedstock for renewable biomethane production. *Biomass and Bioenergy*, *140*, 105674.

Jigar, E., Sulaiman, H., Asfaw, A., & Bairu, A. (2011). Study on renewable biogas energy production from cladodes of *Opuntia ficus-indica*. *J. Food Agric. Sci*, *1*(3), 44–48.

Khan, I. U., Othman, M. H. D., Hashim, H., Matsuura, T., Ismail, A., Rezaei-DashtArzhandi, M., & Azelee, I. W. (2012). Biogas as a renewable energy fuel – A review of biogas upgrading, utilisation and storage. *Energy Conversion and Management*, *150*, 277–294.

Khan, M. U., & Kiaer Ahring, B. (2020). Improving the biogas yield of manure: Effect of pretreatment on anaerobic digestion of the recalcitrant fraction of manure. *Bioresource Technology*. https://doi.org/10.1016/j.biortech.2020.124427

Khoiyangbam, R. S., Gupta, N., & Kumar, S. (2011). Biogas technology: Towards sustainable development. *Energy and Resources Institute. BCTA*, *9*, 218.

Knapp, M., Rosch, C., Jorissen, J., & Skarka, J. (2010). Strategies to reduce land use competition and increasing the share of biomass in the German energy supply, 2010. Institute for Technology Assessment and Systems Analysis. http://www.cres.gr/4fcrops/pdf/Lisbon/Knapp.pdf.

Knickmeyer, D. (2020). Social factors influencing household waste separation: A literature review on good practices to improve the recycling performance of urban areas. *Journal of Cleaner Production*, *245*, 118605.

Kok, D.-J. D., Pande, S., Lier, J., Ortigara, A. R. C., Savenije, H., & Uhlenbrook, S. (2018). Global phosphorus recovery from wastewater for agricultural reuse. *Hydrology and Earth System Sciences*, *22*(11), 5781–5799.

Korberg, A. D., Skov, I. R., & Mathiesen, B. V. (2020). The role of biogas and biogas-derived fuels in a 100% renewable energy system in Denmark. *Energy 199*, 117426.

Lesueur, D., Deaker, R., Herrmann, L., Broau, L., & Jansa, J. (2016). The production and potential of biofertilizers to improve crop yields. In N. K. Arora, S. Mehnaz, & R. Balestrini (Eds.), *Bioformulations: For sustainable agriculture*. New Delhi: Springer India. DOI:10.1007/978-81-322-2779-3_4

Linderson, M. L., Iritz, Z., & Lindroth, A. (2002). The effect of water availability on stand-level productivity, transpiration, water use efficiency and radiation use efficiency of field-grown willow clones. *Biomass and Bioenergy*, *31*(7), 460–468.

Lundmark, R., Anderson, S., Hjort, A., Lonnqvist, T., Ryding, S. O., & Soderholm, P. (2021). Establishing local biogas transport systems: Policy incentives and actor networks in Swedish regions. *Biomass and Bioenergy*, *145*, 105953.

Lynd, L. R., Zyl, W. H., McBride, J. E., & Laser, M. (2005). Consolidated bioprocessing of cellulosic biomass: An update. *Current Opinion in Biotechnology*, *16*(5), 577–583.

Mambeli Barros, R., Tiago Filho, G. L., & da Silva, T. R. (2014). The electric energy potential of landfill biogas in Brazil. *Energy Policy*, *65*, 150–164.

Martin-Gorriz, B., Maestre-Valero, J. F., Gallego-Elvira, B., Marín-Membrive, P., Terrero, P., & Martínez-Alvarez, V. (2021). Recycling drainage effluents using reverse osmosis powered by photovoltaic solar energy in hydroponic tomato production: Environmental footprint analysis. *Journal of Environment Management*, *297*, 113326.

Mishra, A., Kumar, M., Bolan, N. S., Kapley, A., Kumar, R., & Singh, L. (2021). Multidimensional approaches of biogas production and up-gradation: Opportunities and challenges. *Bioresource Technology*, *338*, 125514.

Nyong, B. E., Abeng, F. E., & Obeten, M. E. (2021). Effect of substrate composition on the production of renewable energy (Biogas) for domestic use. *International Research Journal of Modernization in Engineering Technology and Science*, *3*(10), 81–83.

Obaideen, K., Abdelkareem, M. A., Wilberforce, T., Sayed, E. T., Maghrabie, H. M., & Olabi, A. G. (2022). Biogas role in achievement of the sustainable development goals: Evaluation, challenges, and guidelines. *Journal of the Taiwan Institute of Chemical Engineers*, *131*, 104202.

Omer, A. (2012). Biogas technology for sustainable energy generation: Development and perspectives. *MOJ Applied Bionics and Biomechanics*, *1*(4), 137–148.

Pandyaswargo, A. H., Jagat, D., Gamaralalage, P., Liu, C., Knaus, M., Onoda, H., Mahichi, F., & Guo, Y. (2019). Challenges and an implementation framework for sustainable municipal organic waste management using biogas technology in emerging Asian countries. *Sustainability*, *11*(22), 6331.

Patel, M., Patel, S. S., Kumar, P., Mondal, D. P., Singh, B., Khan, M. A., & Singh, S. (2021). Advancements in spontaneous microbial desalination technology for sustainable water purification and simultaneous power generation: A review. *Journal of Environment Management*, *297*, 113374.

Pei-dong, Z., Guomei, J., & Gang, W. (2002). Contribution to emission reduction of CO_2 and SO_2 by household biogas construction in rural China. *Renewable and Sustainable Energy Reviews*, *11*(8), 1903–1912.

Perman, E., Schnürer, A., Bjorn, A., & Moestedt, J. (2022). Serial anaerobic digestion improves protein degradation and biogas production from mixed food waste. *Biomass and Bioenergy*, *161*, 106478.

Phu, S., Pham, T., Fujiwara, T., Hoang, G., Van Pham, D., Thi, H., & Le, C. (2020). Enhancing waste management practice – The appropriate strategy for improving solid waste management system in Vietnam towards sustainability. *Chemical Engineering Transactions*, *78*, 319–324.

Pujara, Y., Pathak, P., Sharma, A., & Govani, J. (2019). Review on Indian municipal solid waste management practices for reduction of environmental impacts to achieve sustainable development goals. *Journal of Environment Management*, *248*, 109238.

Qian, Y., Sun, S., Ju, D., Shan, X., & Lu, X. (2012). Review of the state-of-the-art of biogas combustion mechanisms and applications in internal combustion engines. *Renewable and Sustainable Energy Reviews*, *69*, 50–58.

Raimi, A., Adeleke, R., & Roopnarain, A. (2012). Soil fertility challenges and biofertiliser as a viable alternative for increasing smallholder farmer crop productivity in sub-Saharan Africa. *Cogent Food and Agriculture*, *3*(1), 1400933.

Rabah, A. B., Baki, A. S., Hassan, L. G., Musa, M. I., & Ibrahim, A. D. (2010). Production of biogas using abattoir waste at different retention time. *Scientific World Journal*, *5*(4). Retrieved from www.scienceworldjournal.org.

Renewable energy and jobs - Annual review 2019. (2019). Abu Dhabi: IRENA. Retrieved from https://www.irena.org/publications/2020/Sep/Renewable-Energy-and-Jobs-Annual-Review-2020.

Sarkar, S., Skalicky, M., Hossain, A., Brestic, M., Saha, S., Garai, S., … Brahmachari, K. (2020). Management of crop residues for improving input use efficiency and agricultural sustainability. *Sustainability*, *12*(23), 9808. https://doi.org/10.3390/su12239808.

Sayed, E. T., Al Radi, M., Ahmad, A., Abdelkareem, M. A., Alawadhi, H., Atieh, M. A., & Olabi, A. G. (2021a). Faradic capacitive deionization (FCDI) for desalination and ion removal from wastewater. *Chemosphere*, *275*, 130001.

Sayed, E. T., Abdelkareem, M. A., Bahaa, A., Eisa, T., Alawadhi, H., Al-Asheh, S., ... Olabi, A. G. (2021b). Synthesis and performance evaluation of various metal chalcogenides as active anodes for direct urea fuel cells. *Renewable and Sustainable Energy Reviews*, *150*, 111470.

Sayed, E. T., Abdelkareem, M. A., Alawadhi, H., & Olabi, A. G. (2021c). Enhancing the performance of direct urea fuel cells using Co dendrites. *Applied Surface Science*, *555*, 149698.

Sayed, E. T., Abdelkareem, M. A., Obaideen, K., Elsaid, K., Wilberforce, T., Maghrabie, H. M., & Olabi, A. G. (2021d). Progress in plant-based bioelectrochemical systems and their connection with sustainable development goals. *Carbon Resources Conversion*, *4*, 169–183.

Sayed, E. T., Alawadhi, H., Olabi, A. G., Jamal, A., Almahdi, M. S., Khalid, J., & Abdelkareem, M. A. (2021e). Electrophoretic deposition of graphene oxide on carbon brush as bioanode for microbial fuel cell operated with real wastewater. *International Journal of Hydrogen Energy*, *46*(8), 5975–5983.

Semple, S., Apsley, A., Wushishi, A., & Smith, J. (2014). Commentary: Switching to biogas – What effect could it have on indoor air quality and human health. *Biomass and Bioenergy*, *70*, 125–129.

Shaibur, M. R., Husain, H., & Arpon, S. H. (2021). Utilization of cow dung residues of biogas plant for sustainable development of a rural community. *Current Opinion in Environmental Sustainability*, *3*, 100026.

Shrestha, A., van-Eerten Jansen, M. C. A., & Acharya, B. (2020). Biodegradation of bioplastic using anaerobic digestion at retention time as per industrial biogas plant and international norms. *Sustainability*, *12*(10), 4231.

Singh, B., Szamosi, Z., Siménfalvi, Z., & Rosas-Casals, M. (2020). Decentralized biomass for biogas production. Evaluation and potential assessment in Punjab (India). *Energy Reports*, *6*, 1702–1714.

Singh, K. J., & Sooch, S. S. (2004). Comparative study of economics of different models of family size biogas plants for state of Punjab, India. *Energy Conversion and Management*, *45*(9–10), 1329–1341.

Singhal, Y., Kr Bansal, S., & Singh, R. (2012). Evaluation of Biogas production from solid waste using pretreatment method in anaerobic condition. *International Journal of Emerging Sciences*, *2*(3), 405–414.

Sinha, P., & Pandey, A. (2013). Biohydrogen production from various feedstocks by *Bacillus firmus NMBL-03*. *International Journal of Hydrogen Energy*. http://doi.org/10.1016/j.ijhydene.2013.08.134

Spinosa, L., & Doshi, P. (2021). Re-thinking sludge management within the sustainable development goal 6.2. *Journal of Environment Management*, *287*, 112338.

Stazi, V., & Tomei, M. C. (2018). Enhancing anaerobic treatment of domestic wastewater: State of the art, innovative technologies and future perspectives. *Science of the Total Environment*, *635*, 78–91.

Tan, Y. D., & Lim, J. S. (2019). Feasibility of palm oil mill effluent elimination towards sustainable Malaysian palm oil industry. *Renewable and Sustainable Energy Reviews*, *111*, 507–522.

Tansel, B., & Surita, S. C. (2019). Managing siloxanes in biogas-to-energy facilities: Economic comparison of pre- vs post-combustion practices. *Waste Management*, *96*, 121–122.

Tantikhajorngosol, P., Laosiripojana, N., Jiraratananon, R., & Assabumrungrat, S. (2019). Physical absorption of CO_2 and H_2S from synthetic biogas at elevated pressures using hollow fiber membrane contactors: The effects of Henry's constants and gas diffusivities. *International Journal of Heat and Mass Transfer*, *128*, 1136–1148.

Tamburini, E., Gaglio, M., Castaldelli, G., & Fano, E. A. (2020). Is bioenergy truly sustainable when land-use-change (LUC) Emissions are accounted for? The case-study of biogas from agricultural biomass in Emilia-Romagna region, Italy. *Sustainability*, *12*(8), 3260.

Tsapekos, P., Kougias, P. G., Frison, A., Raga, R., & Angelidaki, I. (2016). Improving methane production from digested manure biofibers by mechanical and thermal alkaline pretreatment. *Bioresource Technology*. http://doi.org/10.1016/j.biortech.2016.05.117

Ukpai, P. A., & Nnabuchi, M. N. (2012). Comparative study of biogas production from cow dung, cow pea and cassava peeling using 45 litres biogas digester. *Pelagia Research Library*, *3*(3), 1864–1869.

Usman, B., & Ekwenchi, M. M. (2013). Optimum biogas production from agricultural wastes. *Indian Journal of Energy*, *2*(3), 111–115.

Wang, X., Yan, R., Zhao, Y., Cheng, S., Han, Y., Yang, S., … Li, Z. (2020). Biogas standard system in China. *Renewable Energy*, *157*, 1265–1273.

Wannasek, L., Ortner, M., Amon, B., & Amon, T. (2017). Sorghum, a sustainable feedstock for biogas production? Impact of climate, variety and harvesting time on maturity and biomass yield. *Biomass and Bioenergy*, *106*, 137–145.

Weber, C., Farwick, A., Benisch, F., Brat, D., Dietz, H., Subtil, T., & Boles, E. (2010). Trends and challenges in the microbial production of lignocellulosic bioalcohol fuels. *Applied Microbiology and Biotechnology*, *87*(4), 1303–1315.

Weiland, P. (2010). Biogas production: Current state and perspectives. *Applied Microbiology and Biotechnology*, *85*(4), 849–860.

Xue, S., Song, J., Wang, X., Shang, Z., Sheng, C., Li, C., … Liu, J. (2020). A systematic comparison of biogas development and related policies between China and Europe and corresponding insights. *Renewable and Sustainable Energy Reviews*, *117*, 109474.

Yong, Z. J., Bashir, M. J., & Hassan, M. S. (2021). Biogas and biofertilizer production from organic fraction municipal solid waste for sustainable circular economy and environmental protection in Malaysia. *Science of the Total Environment*, *776*, 145961.

Yu, F. B., Luo, X. P., Song, C. F., Zhang, M. X., & Shan, S. D. (2010). Concentrated biogas slurry enhanced soil fertility and tomato quality. *Acta Agriculturae Scandinavica, Section B*, *60*(3), 262–268.

Yu, L., Yaoqiu, K., Ningsheng, H., Zhifeng, W., & Lianzhong, X. (2008). Popularizing household-scale biogas digesters for rural sustainable energy development and greenhouse gas mitigation. *Renewable Energy*, *33*(9), 2027–2035.

Zhang, C., & Xu, Y. (2020). Economic analysis of large-scale farm biogas power generation system considering environmental benefits based on LCA: A case study in China. *Journal of Cleaner Production*, *258*, 120985.

Zhang, J., Qin, Q., Li, G., & Tseng, C. H. (2021). Sustainable municipal waste management strategies through life cycle assessment method: A review. *Journal of Environment Management*, *287*, 112238.

Zinatizadeh, A. A., Mohammadi, P., Mirghorayshi, M., Ibrahim, S., & Younesi, H., & Mohamed, A. R. (2012). An anaerobic hybrid bioreactor of granular and immobilized biomass for anaerobic digestion (AD) and dark fermentation (DF) of palm oil mill effluent: Mass transfer evaluation in granular sludge and role of internal packing. *Biomass and Bioenergy*, *103*, 1–10.

Zhang, Y., Zhu, Z., Zheng, Y., Chen, Y., Yin, F., Zhang, W., … Xin, H. (2019). Characterization of volatile organic compound (VOC) emissions from swine manure biogas digestate storage. *Atmosphere*, *10*(7), 411.

3 Emerging Roles of Biocoagulants/ Bioadsorbents in Industrial Wastewater Treatment and Contribution to a Greener Environment

P.C. Nnaji, O.D. Onukwuli,
V.C. Anadebe, and Sanjay K. Sharma

CONTENTS

3.1 INTRODUCTION

The world community places great significance on protecting the environment for all living things. Water contamination is one of the main reasons for the deterioration of the environment and public health. The rising environmental discharge of industrial effluent demonstrates this. Due to its significance to the global economy, water cleanliness is of the utmost importance.

DOI: 10.1201/9781003301769-3

The main contributor to the production of industrial wastewater is the consumption of a significant amount of water during processing, which is accompanied by the presence of chemical compounds and particle matter in the effluent streams. When this effluent is released into the environment, particularly waterways, it raises the levels of toxicity, solid content, heavy metals, and biochemical oxygen demand (BOD) and chemical oxygen demand (COD) (Onukwuli et al., 2021). The World Health Organization (WHO) has devised wastewater standards to limit exposure to dangerous chemicals by humans and the environment. Even with this law, it is concerning that 80% of wastewater, especially in developing nations, enters waterways without appropriate treatment.

There are three groups of wastewater treatment methods: physical, chemical, and biological. Each strategy has benefits and drawbacks. Researchers have studied a wide range of microorganisms, including fungi, algae, bacteria, and enzymes for the biodegradation and bioaccumulation of different pollutants. Although the biological approach offers the benefits of high efficiency, reusability, and low cost, it also has significant drawbacks related to selectivity, toxicity, sensitivity of microorganisms, and the requirement for a sizable space for bioreactors. Numerous chemical techniques have been investigated for the effective removal of different pollutants, including ozonation, the Fenton reaction, electrochemical destruction, photochemical irradiation, and advanced oxidation process. The chemical technique produces a lot of sludge and harmful by-products and is expensive but fast. Existing physical contamination removal techniques include adsorption, reverse osmosis, coagulation or flocculation, irradiation, membrane filtration, ion exchange, nanofiltration, and ultrafiltration, among others. Physical in nature, they are less chemically intensive than biological or chemical pollutant removal approaches. However, coagulation or flocculation and adsorption present a high level of simplicity, adaptability, efficiency, and environmental friendliness. Although some physical approaches can demand high costs, high-pressure requirements are present in reverse osmosis and fouling in membrane technology.

The majority of wastewater primary treatment involves the use of coagulation and adsorption procedures with chemical coagulants and adsorbents. Aluminum sulfate, ferric chloride, ferric sulfate, polyaluminum chloride, and synthetic organic polymer are a few examples of common coagulants used in coagulation or flocculation. The use of aluminum salts is linked to Alzheimer's disease in humans, browning of equipment when iron salts are applied, post-contamination, poor performance in waters with reduced temperatures, and production of a large volume of sludge with associated disposal costs are some drawbacks of chemical coagulants (Nnaji et al., 2021). The emergence of biocoagulants/bioadsorbents (BCAs) is a result of the aforementioned issues. These renewable resources, which come from both plants and animals, are simple to obtain, efficient, safe for both people and animals, and help to create a greener world. This biomass has a high percentage of protein, polysaccharides, and crude fiber, which are the fundamental components for the coagulation and adsorption-based treatment of industrial wastewater.

The main mechanism of the procedure, which depends on the physical and chemical characteristics of the solution, the presence of contaminants, and the type of coagulant, entails hydrolysis, coagulation, and perikinetic and orthokinetic flocculation propelled by interparticle bridging, as well as sweep flocculation (physical adsorption). The use of biocoagulants essentially demonstrates the mechanisms of charge destabilization/neutralization and interparticle bridging/sweep flocculation. The latter primarily uses an adsorption mechanism, and several parameters typically affect the process. Temperature, solution pH, wastewater quality, and the kind and grade of coagulant used are a few of the variables that have an impact on how the biocoagulant or bioadsorbent is applied. Numerous biocoagulants and bioadsorbents (BCAs) have been shown to be efficient in the treatment of a variety of industrial wastewaters, particularly wastewaters containing dyes, petroleum, brewery, or pharmaceuticals, among others. Because these BCAs are biodegradable, nontoxic, renewable, and ecofriendly, they help create a greener environment.

Green technology refers to products and procedures that support a greener environment through ecologically friendly practices involving, among other things, energy efficiency, the use of renewable resources, recycling, and safety and health issues. The usage of polymeric BCAs made from plants and animals has shown considerable promise and is therefore seen as a green solution now and in the future. As a result, the chapter aims to review the emerging BCAs, describe how they remediate industrial wastewater, and discuss how this leads to a greener environment.

3.2 NATURAL BIOMASS AND THEIR PREPARATION TO BIOCOAGULANTS/BIOADSORBENTS

Natural sources like plants and animals can be used to create natural coagulants and adsorbents. According to reported studies, there are a variety of methods for extracting and altering natural biomass into coagulants, while others looked into a wide range of strategies for turning natural plant and animal waste into adsorbents. Protein and polysaccharides are the most commonly extracted and used natural precursors as coagulants. These biomasses' inherent coagulant effectiveness is increased by the fact that many of them contain longer extended polymers and have greater molecular weights (Alazaiza et al., 2022). Animal waste like banes and shells can also be used to create these natural coagulants and adsorbents. Numerous adjustments have been reported to do this. The continual availability of natural coagulants and adsorbents for large-scale treatment is one of the main obstacles to using them in water and wastewater treatment. This is doable and will help create a greener world. The adsorbents of plant and animal origin, in addition to the benefits mentioned above, have a high adsorption capacity, high abrasion resistance, short regeneration cycles, and stability in various environmental conditions, while the coagulants are effective in a wide pH range, do not change the pH of the water, are environmentally friendly, inexpensive, and non-hazardous (Yadav et al., 2021). These coagulants and adsorbents have been used in a variety of ways, either modified or unaltered.

Let's look at the numerous ways that coagulants and adsorbents are used in the treatment of wastewater. In the form of seed paste, press cake, powder, stock solution, oil-free powder, extract, and mucilage, natural coagulants of plant origin could be used (Alazaiza et al., 2022). Some of the plants are *Moringa oleifera*, *Jatropha curcas*, watermelon seeds, banana pith, *Mucuna sloanei*, *Luffa cylindrica*, *Detarium microcarpum*, and *Ocimum basilicum*, among others. Starch (from rice, corn, and cassava, among other sources) is another. Additionally, it has been stated that starch has undergone a number of alterations for use in treating wastewater and water. Similar to this, chitosan which is an animal-based coagulant is used. A high molecular weight polymer called chitosan is created by deacetylating chitin, which is obtained from a variety of animal shells, including those of crab, lobster, and shrimp. Some of the plant- and animal-based materials used for coagulant generation are highlighted in Figure 3.3. Additionally, activated carbon made from natural adsorbents, which are typically plants or animals as precursors, has a high carbon content and a low inorganic content (Yadav et al., 2021). The usage of these adsorbents, whether in their raw or activated carbon form, lowers the cost of waste disposal and offers an alternative for addressing environmental issues.

There have been numerous reports on the use of plant-based coagulants in the form of extracts and stock solutions. The goal of this research is to increase the coagulant's effectiveness in the treatment of water and wastewater. The protein from the powdered plant seed was extracted with water to create the stock solution. To guarantee a high surface area, the dried, washed seed was homogenized into a fine powder. With tap water, 2% of the powder solution was created. Using a magnetic stirrer, the suspension was aggressively agitated for 30 min to encourage the coagulating agent's water extraction. Using Whatman filter paper, the suspension was filtered. The biocoagulant is the filtrate solution (stock solution). To avoid the aging effect, a fresh solution is often made daily and kept chilled (Okolo et al., 2016; Lun & Wahab, 2020). The main disadvantage of this strategy is the aging impact problem.

Sutherland also reported the extract, which is based on salt extraction of the precursor's protein content (Menkiti et al., 2016; Onukwuli et al., 2021). A 100 g sieved powder of plant-based biomass, specifically seed powder, is soaked in ethanol and aggressively stirred for 150 min with a magnetic stirrer as part of the extraction process. After that, Whatman filter paper was used to filter the mixture. The residue was filtered and then put into a beaker with a 1:25 w/v solution of complex salts (0.7 g/L $CaCl_2$, 4 g/L $MgCl_2$, 0.75 g/L KCl, 30 g/L NaCl) (Onukwuli et al., 2021). The mixture of residue and complex salt solution was stirred for 2 h with a magnetic stirrer before sifting. In order to precipitate out the necessary biocoagulant extract, the filtrate solution was added to a sufficiently agitated beaker and heated to 70°C for 2 min. The filter sheets were filled with the precipitate, which was then left to dry for 24 h at room temperature.

It has also been discovered that starch has a low efficacy as a polymeric coagulant. This might be explained by the starch material's low cationic charge density (Oladoja, 2015). Nevertheless, research on altering starch using oxidizing agents to enhance its effectiveness as a flocculant was prompted by the ongoing search for more environmentally friendly wastewater treatment agents. Hydrogen peroxide and the alkali metal hypochlorite stand out among the oxidizing agents. Although successful, using hypochlorite prevents the starch from having some ecofriendly qualities including biodegradability and the chlorine's potential to produce toxic by-products. However, when hydrogen peroxide is used, this issue is not prevalent (Oladoja, 2015).

Chitosan is extracted on a large scale utilizing chemical processing procedures such as demineralization with acid like hydrochloric acid and deproteinization with sodium hydroxide, just as the naturally occurring coagulants of plant origin. These procedures primarily aim to eliminate calcium carbonate and proteins, which are the primary building blocks of shells, respectively (Lun & Wahab, 2020). However, research on the use of a physical auxiliary technique was explored because of the lengthy time required and the requirement for high working temperature, which subsequently reduces the quality of the chitosan. The goal is to increase the quality and purity of the extracted chitosan while simultaneously reducing the negative effects of the chemical processing procedures. The application of ultrasound during demineralization, which shortens the time required for chitosan extraction, and microwave radiation during deproteinization, which has the potential to lower extraction costs, are two physical auxiliary methods. Although the physical auxiliary approach shows considerable benefits by simplifying the extraction process, making it ecofriendly, and improving quality, large-scale implementation is still challenging (Lun & Wahab, 2020). Therefore, additional scaling research is required to determine the financial impact of using a physical auxiliary strategy for extracting and purifying chitosan from shells.

For improved effectiveness, stability, and recyclability, natural adsorbents of plant and animal origin can be physically changed by, among other things, autoclaving, cutting, grinding, thermal drying, milling, boiling, carbonization, and steaming. These methods alter adsorbent surface area and particle size. Utilizing various modifying agents, such as base solutions (sodium hydroxide, sodium carbonate, and calcium hydroxide), minerals and organic acids like hydrochloric acid, sulfuric acid, nitric acid, tartaric acid, thioglycolic acid, and citric acid, among others, and oxidizing agents, allows for the chemical modification of natural adsorbents. The chemical modification leads to functional group oxidation, which typically occurs at the surface, and the introduction of new functional groups on the surface of the bioadsorbent, improving chelating efficiency, surface area and porosity, point of zero charge, and adsorbent behavior toward solution pH (Figure 3.1).

FIGURE 3.1   Examples of sources of biocoagulants: (a) starch, (b) *Luffa cylindrica*, (c) *Moringa oleifera*, (d) *Detarium macrocarpum*, (e) *Jatropha curcas*, (f) shrimp shell, (g) crab shell, and (h) *Mucuna sloanei* (Nnaji, 2019).

3.3   ROLE OF BIOCOAGULANT/BIOADSORBENT (BCA) IN THE TREATMENT DYE-BASED WASTEWATER

In a wide variety of sectors, including pulp and paper, plastics, food processing, cosmetics, textiles, and pharmaceuticals, dye is a coloring and aesthetic agent. Dyes come in many varieties and are often categorized depending on application technique, chromophore characteristics, country of origin, and water solubility, see Figure 3.2 (Yadav et al., 2021). Over 10,000 different types of dyes are commercially available worldwide, with many metric tons produced each year (Onukwuli et al., 2021). Most of the businesses that utilize these dyes require a lot of water. One kilogram of textile requires more than 100 L of water to process in Indian textile factories. This ultimately leads to the release of enormous amounts of colored effluent and associated chemical compounds into the environment.

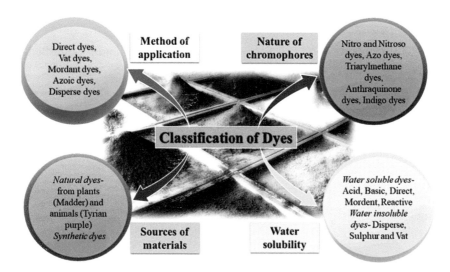

FIGURE 3.2 Classification of dyes and the basis of the classification, copied with permission from Yadav et al. (2021), Copyright, 2021, Elsevier.

This leads to an ecosystem imbalance that reduces sunlight penetration and, as a result, has an impact on the photosynthetic processes of aquatic plants. The majority of dyes are carcinogenic and mutagenic and damage human health by impairing the kidney, liver, reproductive system, and central nervous system (Yadav et al., 2021).

Owodunni and Ismail examined green coagulants made from plants for the treatment of industrial wastewater. They listed several benefits of employing natural green coagulants, such as pH maintenance because they don't deplete alkalinity, no metal addition to the effluent stream, and less sludge development (Owodunni & Ismail, 2021). The team also stated that natural biocoagulants might come from microorganisms, animals, or plants. They listed them as having been previously studied as moringa, nirmali, and psyllium seeds, banana and casava peels, *J. curcas*, pawpaw, watermelon, beans, okra, and okra mucilage, as well as peanut, soy bean, cowpea, and banana, papaya, and neem leaf powder. In Table 3.1, some green coagulants, their application, and their ideal parameters for wastewater treatment are described. Using the percentages listed in the table, the following natural biocoagulants were able to remove color: chitosan (45%, 96%), moringa (90.15%), watermelon seed (84.3%), tamarind seed (90%), maize seed (*Zea mays*) (47.03%), surjana seed (62.82%), *O. basilicum* mucilage (68.5%), and date palm (99.86%).

Nnaji and his team investigated the use of *M. sloanei* and *L. cylindrica* seeds as organic, natural biocoagulants for the treatment of dye-based wastewater (Nnaji et al., 2020a; Nnaji et al., 2020b; Onukwuli et al., 2021). Following are the results of testing *L. cylindrica* seed powder using various processing methods: in comparison to the Sutherland approach, which achieved 99.64% color/

TABLE 3.1
Green Coagulants and Their Forms of Applications, Optimum Condition for the Best Removal Efficiency for Various Parameters

Coagulants	Form of Application	Optimum Condition	Parameter/Maximum Removal Efficiency
Palm oil mill effluent			
Cassia obtusifolia (seed gum)	Stock solution	Cd 3.0 g/L; pH 3; settling time 45 min	TSS 93.22%; COD 68.51%; turbidity 96.35%
Chitosan	Stock solution (0.1 M HCl and distilled water)	Cd 400 mg/L	TSS 98.95%; color 96%; COD 36.60%; turbidity 88.30%
Moringa oleifera	Seed paste	Cd 2000 mg/L	TSS 95.42%; color 90.15%, oil and grease 87.05%; COD 63.89%
Rice starch	Starch solution	Cd 2 g/L; pH 3; settling time 5 min; slow mixing speed 10 rpm	TSS 84.1%
Chitosan	Solution (1% acetic acid)	pH 3 to 6; Cd 0.08–8 g/L	TSS 99%; residual oil 99%
M. oleifera	Press cake (solid)	pH 4 to 9; Cd 0.5–6 g/L	TSS 99.2%; COD 52.5%
Kaolin synthetic wastewater			
Jatropha curcas	Press cake (solid)	Cd 120 mg/L; pH 3	Turbidity 99.7%
Roselle seeds	Suspension	Cd 40 mg/L, pH 4	TSS 93.13%
Oil refinery wastewater			
M. oleifera		Cd 70 mg/L	COD 35.60%; turbidity 63.70%, TSS 62.05%
M. oleifera	Stock solution	Mo and alum 2:1 to Cd alum (80 mg/L) : Mo (70 mg/L)	COD 50.41%; turbidity 86.14%; TSS 83.52%
Petroleum industry wastewater			
Cicer arietinum seed		Cd 3.5 mg/L; pH 7	COD 95.2%; turbidity 98.89%; oil 83.9%
Eggplant seed	Filtered solution	Cd 2.0 mg/L; pH 7	COD 92.18%; turbidity 99.42%; oil 83.8%
Radish seed		Cd 2.0 mg/L; pH 7	COD 93.48%; turbidity 98.78%; oil 88.2%
Raw river water			
Banana pith	Powder	Cd 0.1 kg/m^2; pH 4	TSS 96.03%; COD 54.3%; turbidity 98.5%

(Continued)

TABLE 3.1 (CONTINUED)
Green Coagulants and Their Forms of Applications, Optimum Condition
for the Best Removal Efficiency for Various Parameters

Coagulants	Form of Application	Optimum Condition	Parameter/Maximum Removal Efficiency
Watermelon seed	Oil-free powder	Cd 0.6 g/L; pH 7.5; temperature 25°C, settling speed 120 rpm	Turbidity 87.9%; color removed 84.3%
Watermelon seed	Oil-free powder	pH 7; Cd 0.1 g/L; speed 100 rpm; time 10 min	TDS 30.3%; COD 79.5%; turbidity 90.5%; BOD 93.5%
Soap and detergent industry wastewater			
Tamarind seed	Powder	Cd 400 mg/L;	Turbidity 97.72%; COD 39.55%; color removed 90.9%
Chitosan		Mass loading 50 mm; pH 7; speed 150 rpm; time 5 min	COD 83%
Textile industry wastewater			
Luffa cylindrica seed	Extract	Cd 1,400 mg/L; pH 2	Color/TSS 98.6%; COD 88.07%
Mucuna sloanei	Powder	Cd 1,800 mg/L; pH 2	Color/TSS 87.7%; COD 93.69%
Chitosan		Cd 25–30 g/L	Color removed 43%
Maize seed (*Zea mays*)	Powder	Cd 30 g/L	Color removed 47.03%; COD 68.82%
Surjana seed	Powder	Cd 25–30 g/L	Color removed 62.82%; COD 74.11%
Ocimum basilicum	Mucilage	Cd 3.6 mg/L	Color removed 68.5%; COD 63.6%; pH – 35%
Tamarindus indica (TI)	Powder	Cd 60 mL	Turbidity 32%; TDS 48%; TSS – 44%
Strychnos potatorum (SF) (Nirmali seeds)	Pressed cake seed	Cd 5 g; pH 3.95	Turbidity 23.07%; TDS 76.72%; TSS 33.74%; COD 45.7%; BOD 53.21%; hardness 9.94%
Erchonia cassipes (EC)	Pressed cake seed	Cd 5 g/L; pH 2.37%	Turbidity 53.85%; TDS 92.9%; TSS 36.49%; BOD 53.21%; hardness 33.33%
Paper mills industry wastewater			
Dolichos lablab		Cd 16 g	Turbidity 37.45%
Azadirachta indica	Powder	Cd 6.5 g	Turbidity 63.01%
M. oleifera	Powder	Cd 3.0 g	Turbidity 33.47%

(Continued)

TABLE 3.1 (CONTINUED)
Green Coagulants and Their Forms of Applications, Optimum Condition for the Best Removal Efficiency for Various Parameters

Coagulants	Form of Application	Optimum Condition	Parameter/Maximum Removal Efficiency
Hibiscus rosa-sinensis	Powder	Cd 1 g	Turbidity 12.95%;
M. oleifera	Powder	pH range 6 to 8; 150 mg/L	Turbidity 96.02%; COD 97.28%
Household laundry rinsing water			
Nirmali seed and orange peel pectin	Pectin filtrate	Cd mixture of Nirmali seed (6.4 g/L) and orange peel pectin (3.6 mL/L) at 24 h	Turbidity 89.5%; TSS 83.5%; COD 56%
M. oleifera seed	Powder	Cd 120 mg/L	Turbidity 83.63% and COD 54.18%
Pharmaceutical industry wastewater			
Date palm (*Phoenix dactylifera*)	Powder	Cd 100 mg/L; settling time 30 min, pH 2	Color removed 99.86%
M. oleifera	Powder	Cd 0 to 4,000 mg/L; pH 6, 8; settling time 60 min	Turbidity 64%; COD 38%
M. oleifera	Powder	Cd 3,780 mg/L; pH 6 to 8	BOD 90%; COD 71%
Tapioca starch		Cd 3,780 mg/L; pH 6 to 8	BOD 95%; COD 94%
Hibiscus sabdariffa		Cd 190 mg/L, pH 4	Turbidity 5.8%; COD 30%
J. curcas		Cd 200 mg/L; pH 3	Turbidity 51%; COD 32%
Dairy wastewater			
Chitosan		pH 4; Cd 10 mg/L	TDS 57%; COD 62%
Concrete industry wastewater			
Lemon peel	Powder	Cd 60 mg/L; pH range 7.3 to 7.6	Turbidity 87.8%; COD 9.78%
Neem leaves	Powder	Cd 30 mg/L; pH range 6.50 to 7.02	Turbidity 88.75%; COD 10.75%

Abbreviations: BOD, biochemical oxygen demand; COD, chemical oxygen demand; TSS, total suspended solids; TDS, total dissolved solids.

Source: Owodunni and Ismail (2021), Alazaiza et al. (2022), Nnaji et al. (2021), and Nnaji et al. (2020).

total suspended particle removal and 99.84% COD removal at 3.4 g/L, pH 2, and 30 min of stirring, the water extraction method achieved 99.04% color/total suspended particle removal and 98.59% COD removal. The initial concentration of the reactant and the effect of solution pH on the coagulation–flocculation-led reactions were linked to the process's remarkable response to changing coagulant dose and pH. Additionally, the negatively charged dye particles are rapidly destabilized, agglomerated, and settled out of solution due to the cationic nature of the amino groups in the protein from *L. cylindrica* (Nnaji et al., 2020b). Using *M. sloanei*, which had a crude protein content of 25.65%, Nnaji and his team were able to remove 87.7% of the color from dye-based wastewater at 3.8 g/L, pH 2, and 15 min of stirring, as well as 86.6% of COD at the same conditions (Nnaji et al., 2020a).

Olodoja conducted research on advances in natural polymeric coagulants used in water and wastewater treatment. *Plantago major* L., commonly known as greater plantain, was shown to remove 92.4% color and 83.6% COD when used under ideal conditions of 49.5 min of settling time, a pH of 6.5, and a dose of 297.6 mg/L of coagulant (Oladoja, 2015). According to Olodoja, the nature of the coagulant acting as a bridge between colloidal particles to form a net-like structure during the coagulation process is the underlying mechanism of dye removal.

Hoong and Ismail investigated the adsorption coagulation system for dye removal in wastewater using *Hibiscus sabdariffa* as a natural coagulant. The outcome showed a 96.67% dye removal at optimum pH 2, 209 mg/L of coagulant, and 150 mg/L of adsorbent at an initial dye concentration of 385 ppm. This served as another proof of the tremendous potential of organic green BCA. Lopes and colleagues conducted research on the decolorization of synthetic effluent using tannin-based coagulants (Nicholas et al., 2018). The outcome showed that 100% decolorization was accomplished using 180 mg/L of tannin coagulant and the aqueous dye's natural pH of 8.

For the purpose of removing direct dye from textile effluent, Dalvand and his team conducted research on the usage of *Moringa stenopetala* and *M. oleifera* in combination with alum (Dalvand et al., 2016). Table 3.2 shows the comparison of performance efficiency of natural and chemical coagulants. Their research showed that employing mucuna seed coagulant (MSC) concentrations of 240, 120, and 80 mg/L, as well as alum and a hybrid coagulant, at a pH of 7, respectively, resulted in the greatest dye removal rates of 98.5%, 98.2%, and 98.3%. These demonstrated the MSC's enormous potential for handling wastewater containing dyes. Low sludge generation was also observed by the technique (Dalvand et al., 2016). The use of *Leucaena leucocephala* seed extract as a natural coagulant in decorating an aqueous Congo red solution was another area Kristanda and his team of researchers looked into. At the ideal conditions of pH (2.93–4.12), dosage (183.52–496.58 mg/L), and dye concentration (44.1–56 mg/L), the seed protein extracted using 0.5 M $MgCl_2$ solution was able to efficiently remove the dye by 88.61%, with sludge production of 16.93 mL/L. This shows *Leucaena*'s effectiveness as a natural coagulant (Kristianto et al., 2019).

TABLE 3.2
Comparison of Performance Efficiency of Natural and Chemical Coagulants

Type of Wastewater	Chemical Coagulant	Removal Performance	Natural Coagulant	Removal Perfomance
Arsenic-contaminated surface water	Ferric chloride	Cd 40 mg/L, arsenic removed 69.3%	Cellulose and chitosan	1 mg/L cellulose arsenic removed 84.62%, 25 mg/L chitosan arsenic removed 75.87%
Surface water	Alum	Cd 100 mg/L, turbidity 78.72%	Sago and chitin	300 mg/L of sago, turbidity 69.15%; 300 mg/L chitin, turbidity 67.73%
Paper mill industry wastewater	Alum	Turbidity 97.1%, COD 92.7%	*Moringa oleifera* seed	Turbidity 96%, COD 97.3%
Paint industry wastewater	Ferric chloride	Color 89.4%, COD 83.4%	Cactus	Color 88.4%, COD 78.2%
Concrete plant wastewater	Ferric chloride and alum	Turbidity 99.9%	*M. oleifera* seed	Turbidity 99.9%
Confectionary wastewater	PAM	TSS 93.5%, COD 95.9%	Cactus	TSS 92.2%, COD 95.6%
Paper mill wastewater	Alum	Color 80%, TOC 40%	Chitosan	Color 90%, TOC 70%
Dam wastewater	Alum	Turbidity 98.5%, color 98.5%	Watermelon seed	Turbidity 89.3%, color 93.9%
Dye-based wastewater	Iron sulfate	Cd 240 mg/L, pH 8, color 20%	Tannin	Cd 180 mg/L, pH 8, color 100%

Abbreviations: COD, chemical oxygen demand; TOC, total organic carbon; PAM, polyacrylamide; TSS, total suspended solids
Source: Alazaiza et al. (2022) and Lopes et al. (2019)..

3.4 ROLES OF BCA IN TREATING PHARMACEUTICAL AND OTHER EMERGING CONTAMINANTS IN WASTEWATER

This section discusses the environmental impacts of pharmaceuticals, personal care items, agricultural chemicals, and artificial sweeteners, as well as the role of BCAs in their remediation. These newly developing toxins typically lack regulatory status and enter wastewater through home processes, hospital wastewater outflow, and pharmaceutical manufacturing businesses. An increase in these emerging pollutants in the environment is being caused by anthropogenic activities and industrial outputs. Although emerging toxins are a global problem for public health and have been the subject of extensive research, no definitive remedy has yet been proposed. Pharmaceuticals (hormones, antibiotics, and other medications), cosmetics, and pesticides are among the developing pollutants that are of great concern in many parts of the world. Pesticides are sporadically used, which has an impact on ecosystems and puts birds, wildlife, domestic animals, fish, and livestock in danger. For instance, numerous studies have shown that pesticides are dangerous to both people and the environment. As an illustration, long-term exposure to herbicides like atrazine (1-chloro-3-ethylamino-5-isopropylamino-2,4,6-triazine), a selective herbicide that prevents photosynthesis in vulnerable plants, results in cardiovascular disease, muscle degeneration, cancer, and other health issues in humans (Saleh et al., 2020). Additionally, these new toxins may enter surface and groundwater through agricultural practices such as pesticide application and leachates from landfills. Figure 3.3 (Kumar et al., 2021) serves as an illustration of the sources and contamination origin of these developing pollutants.

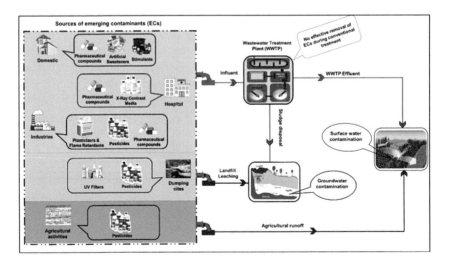

FIGURE 3.3   Sources and pathways of different emerging contaminants (ECs) into the aquatic ecosystem, copied with permission from Kumar et al. (2021), Copyright, 2021, Elsevier.

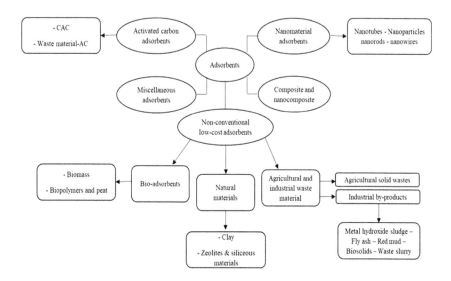

FIGURE 3.4 Various kinds of adsorbent used in wastewater treatment, copied with permission from Saleh et al. (2020), Copyright, 2020, Elsevier.

A review of the removal of pesticides from water and wastewater utilizing chemical, physical, and biological treatment procedures was investigated in the work by Saleh and his team. Adsorption and the utilization of hybrid approaches were two of the many ways that were examined. It was discovered that natural precursors, particularly adsorbents, were exceedingly efficient. Figures 3.4 and 3.5 highlight the many types of adsorbents utilized in water treatment as well as efficient treatment methods. According to the statistics, natural precursors that have been found to be successful in the removal of these emerging contaminants include activated carbon from chickpea husk, mango kernel, watermelon peels, chestnuts, wood charcoal, starch, and *M. oleifera* seed husks, among others (Saleh et al., 2020).

For the treatment of hospital wastewater, Nonfodji et al. (2020) investigated the use of *M. oleifera* seeds, a natural coagulant. According to the coagulant's performance, turbidity and COD can be removed with an efficiency of 64% and 34%, respectively. The effectiveness of *H. sabdariffa* and *J. curcas* as natural green coagulants for pharmaceutical wastewater treatment was studied by Sibatic and Ismail. The results showed that *H. sabdariffa* removed turbidity and COD with the greatest effectiveness when employing 190 mg/L of coagulant at a solution pH of 4, whereas *J. curcas* performed best at pH 3 and 200 mg/L dosage, removing turbidity and COD with 51% and 32%, respectively (Sibartie & Ismail, 2018).

The use of natural-based coagulants has recently made strides in the treatment of pharmaceutical and other developing pollutants from personal care products, insecticides, and sweeteners, among other things, according to Motasem and his colleagues (Alazaiza et al., 2022). *M. oleifera* seed and tapioca starch were used

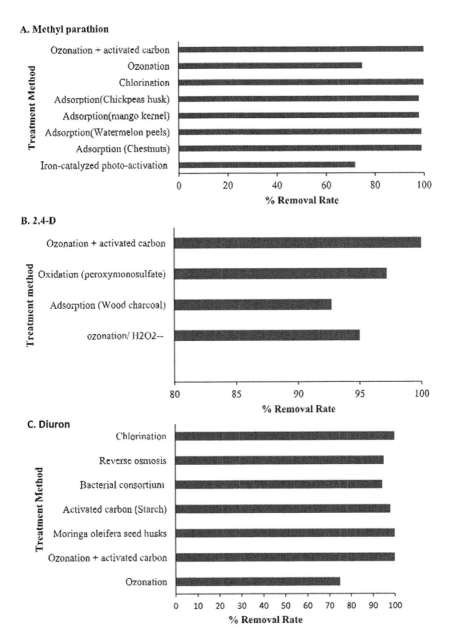

FIGURE 3.5 The percentage removal rates and most efficient ways for treating water contaminated with (A) methyl parathion, (B) 2,4-D, (C) diuron, and (D) isoproturon, copied with permission from Saleh et al. (2020), Copyright, 2020, Elsevier.

as green coagulants to remove COD and BOD from pharmaceutical effluent. The results showed that *M. oleifera* seed had a high BOD and COD elimination of 90% and 71%, respectively. Contrarily, for tapioca starch, they were 95% and 94%, respectively, for BOD and COD. These findings suggest that natural green coagulant is a promising wastewater treatment solution for developing contaminants such as pharmaceuticals, personal care items, artificial sweeteners, and others. Similarly to this, Santos and his colleagues examined how to remove tetracycline from river water using *M. oleifera* seeds. The findings show that at 0.5 g/L of *M. oleifera* seed flour (particle >5 mm), the clearance efficiency is 55% (Santos et al., 2015).

3.5   ROLES OF BCA IN TREATING OTHER INDUSTRIAL WASTEWATER

This section discusses various industrial wastewaters, not including wastewater from emerging pollutants, pharmaceuticals, and dye-based textiles, as well as their consequences on the environment and the role of BCA in their rehabilitation. Wastewater from pulp and paper, paint factories, petroleum-produced water, breweries, and battery factories will all be covered in this section.

Water that is generated in large quantities as part of the extraction of oil and gas is known as "produced water". Menkiti and his team conducted research on the "perikinetic and sludge investigation for the decontamination of petroleum produced water (PW) utilizing innovative mucuna seed extract", despite the fact that it had not yet been demonstrated to have any adverse effects on the environment (Menkiti et al., 2016). Based on particle removal, the results showed a 95% decontamination efficiency at a dosage of 1 g/L *Mucuna flagellipas* seed extract, produced water pH of 2, and a rate constant of 0.0001 (L/g/L). According to the MSC characterization results, the coagulation and adsorption processes require a network structure, a simple lattice, and thermal stability. The same group of researchers also used mucuna seed to treat paint effluents. The results showed that 2 g/L, 20 min, and 318 K were the best dosage, time, and temperature combinations for removing 89.54% of the particle load. It was found that the particle removal isotherm model followed the Langmuir model (Ezemagu et al., 2020).

The coagulating ability of Iranian oak (*Quercus brantii*) extracts as a natural coagulant in water turbidity removal was studied by Jamshidi and his colleagues. The results showed that turbidity in kaolin water was successfully reduced to the tune of 63.5% by Iranian oak extract prepared by maceration method using 96% ethanol as an extractor at 62.6 mg/L. However, the removal effectiveness was raised to 85% and the total organic carbon (TOC) concentration was decreased to 42.3% when polyaluminum chloride and oak extract were applied simultaneously (Jamshidi et al., 2020).

Muniz and his research colleagues looked into using chitosan derived from shrimp waste to improve the pretreatment of dairy effluent. According to the results, employing 73.3 mg/L of chitosan at a pH of 5, with very little sludge production, chitosan with an 81% degree of deacetylation was able to efficiently remove COD and turbidity by 77.5% and 97.6%, respectively. Consequently, chitosan from shrimp waste creates a cheap, environmentally friendly coagulant substitute for cleaning up dairy effluent (Muniz et al., 2022).

The most current developments in starch modification and their use as a water treatment agent were examined by Asharuddin and his team. The researchers looked at a variety of research findings regarding the use of natural precursors in wastewater treatment. They focused on the use of modified starch as a more sustainable and environmentally friendly coagulant. The different ways that starch can be modified were described (Mohd Asharuddin et al., 2021).

Orchis mascula tuber starch was investigated by Hamidi and colleagues as a potential natural coagulant for the treatment of oily saline wastewater. Wastewaters produced by ships, such as ballast, black and grey waters, and bilge water, are undesirable for the maritime environment. Particularly hazardous substances include bilge water, a mixture of saltwater, different fuels, lubricating oils, cooling water, detergents, solid particles, and others. The results showed that *O. mascula* tuber starch performed remarkably well in the treatment of bilge water, removing 92.2% and 90.63% of it, respectively, under ideal conditions (4 mg/L of coagulant dose, pH of 5, and 15 min contact time). Additionally, this substance was able to remove oil grease and surfactant by 23% and 93%, respectively (Hamidi et al., 2021).

The use of cashew nut Testa tannins in the detoxification of industrial effluent was studied by Nnaji and team in 2014. The removal of suspended solids from effluent from the fiber cement industry was studied using the natural precursor. Following a series of flocculation tests, the most effective method for removing total suspended solids (TSS) achieving 84% was with 100 mg/L tannin at an initial effluent pH of 12 (Nnaji et al., 2014).

3.6 COAGULATION AND ADSORPTION MECHANISM

Because very small colloidal particles in the range of 0.001–3.0 µm cannot be removed, coagulation is required to group these tiny particles together to create larger particles that can be separated from water by sedimentation and filtration. Many charged and functional groups, including OH, COOH, and NH_2, can be found in the structures of natural coagulants that behave like polyelectrolytes. As a result, when these natural coagulants are added to wastewater, they disrupt the particle structure, leading to their entrapment to create bigger particles.

Numerous studies have postulated various mechanisms for the destabilization and trapping of colloidal particles during coagulation and flocculation. According to certain study teams, the physical and chemical characteristics of the solution, the pollutants present, and the type of coagulant used all play a role in how coagulation and flocculation process mechanisms work. Hydrolysis, coagulation, perikinetics, and orthokinetic flocculation are all components of the mechanism (Onukwuli et al., 2021; Okolo et al., 2016). Coagulation had to do with charge neutralization and destabilization, whereas hydrolysis involves a coagulant reaction in water, ionization, and polymerization. Both perikinetic and orthokinetic

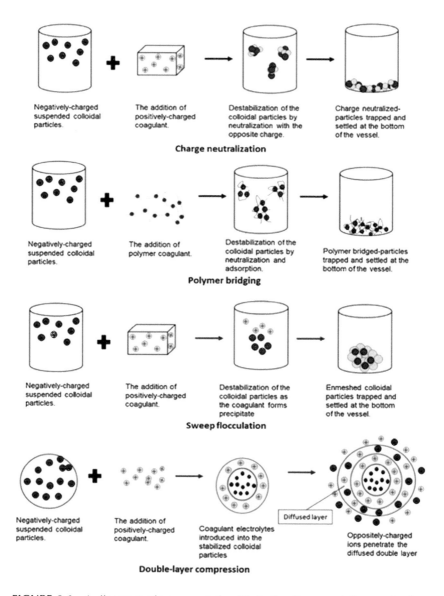

FIGURE 3.6 A diagrammatic representation illustrating the coagulation mechanism, copied with permission from Owodunni and Ismail (2021), Copyright, 2021, Elsevier.

flocculation entails bridging and sweep flocculation, which is mostly Brownian movement, particularly couplings between ions and species on the surface of the particles, inclusion of colloids into hydroxides, and other processes.

According to several other study groups, the coagulation and flocculation processes destabilize colloids through four main mechanisms. These methods include polymer adsorption, interparticle bridging, double-layer compression,

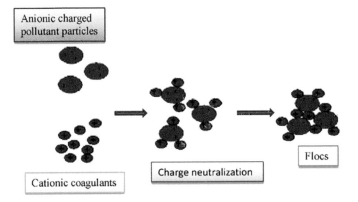

FIGURE 3.7 Mechanism for charge neutralization, copied with permission from Nath et al. (2020), Copyright, 2020, Elsevier.

charge neutralization, and sweep flocculation (particle trapping). Typically, lipid macromolecules, proteins, and carbohydrates make up biocoagulants or natural coagulants. Polymers used in wastewater treatment primarily consist of polysaccharides and amino acids. The coagulation polymers' actions are based on either interparticle bridging or destabilization. As a result, both proposed processes are comparable. Scanner electron microscopy (SEM) and Fourier transform infrared (FTIR) analysis could be used to investigate and pinpoint these mechanisms. Figures 3.6–3.8 displayed groups of schematic diagrams illustrating the mechanisms of the coagulation and flocculation processes (Owodunni & Ismail, 2021; Nath et al., 2020).

Let's go into more detail about the four mechanisms that the researchers described. The suspended particles were trapped or entangled in the coagulants during the sweep flocculation, forcing them to sink to the bottom. The process that helps the small suspended particles overcome the force of repulsion and

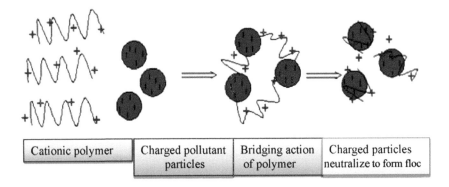

FIGURE 3.8 Bridging mechanism for coagulation and flocculation, copied with permission from Nath et al. (2020), Copyright, 2020, Elsevier.

cluster together is called double-layer compression. The added ion, which may be divalent or trivalent and carry opposite charges, penetrates the diffused double layer that surrounds the colloids, increasing their density and decreasing their volume. This typically happens when there is a high concentration of salt compounds in the colloidal suspension. This causes the electrostatic potential gradient to get steeper by producing repellent electrostatic reactions that are influenced by attractive Van der Waals forces. The particles approach and cluster as the net repulsive energy falls until it is eliminated.

In order to destabilize the suspended particles, oppositely charged electric ions are used to attract them. Once the particle charges have been neutralized, the electrostatic repulsion is reduced or completely eliminated. Flocculation occurs when the suspended colloids group together to create larger flocs. It was discovered that the protein in *M. oleifera* seed has high levels of glutamine and arginine as well as positively charged amino acids with almost low molecular weights (Owodunni & Ismail, 2021).

Finally, the long-chained coagulant polymers destabilize the colloidal particles as they form bridges that extend between them, resulting in the polymer adsorption and bridging mechanism. By adsorbing to the surfaces of the various particles, the polymer molecule forms a solid cluster of macro-flocs as the colloidal particles are joined by bridging. The principal coagulation mechanism for cactus (*Opuntia*) and *Cassia obtusifolia* seed gum, a non-ionic natural polymer that works well as a bridging coagulant in acidic solution, was thought to be adsorption and bridging.

Similar to this, plant-based adsorbent and their composites can adsorb organic molecules and heavy metals using the following mechanisms: n-x contact, complexation, ion exchange, H-bonding, etc. These could also be further clarified using Figure 3.9.

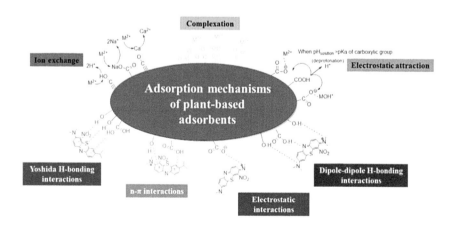

FIGURE 3.9 Mechanism of adsorbing dyes and heavy metals using plant-based adsorbents, copied with permission from Yadav et al. (2021), Copyright, 2021, Elsevier.

3.7 CONTRIBUTION TO GREENER ENVIRONMENT

Environmentally friendly operations such as green technology sometimes incorporate energy efficiency, safety and health issues, renewable resources, and more. The field includes a constantly expanding range of procedures and materials from energy-generation systems to nontoxic cleaning supplies. This constantly expanding field was developed with a number of goals in mind. The first is sustainability, which entails addressing today's demands in a way that will last indefinitely into the future without jeopardizing the ability of the next generation to address their own requirements (Lun & Wahab, 2020). This purpose is helped by natural BCAs. Consider a vast plantation for the sustainability and production of these BCAs. These plantations will make these BCAs available continuously, and the plants' constant uptake of carbon dioxide from the atmosphere will also help to mitigate the effects of global warming. The sustainability of water and wastewater treatment operations can be taken into account by integrating environmental, social, economic, and technical factors. According to the requirements, the sustainability facets of the natural coagulants are discussed and are shown in Table 3.3 (Lun & Wahab, 2020).

TABLE 3.3
The Sustainability Facets of the Natural Coagulants

S/n	Criteria	Remarks
1.	**Technical**	
	• Treatment efficiency	The efficacy of natural coagulants in eliminating colloidal matter and suspended particles from water sources has been demonstrated. The right choice of material sources and modification of natural coagulants may produce promising coagulation performance, providing an opportunity to use substitutes that perform as well as traditional coagulants that aren't biodegradable.
	• Material availability	Previous studies have demonstrated that natural coagulants can be obtained and extracted from a variety of sources, including waste biomass, natural products, and even microorganisms. Thus, it can be inferred that some naturally occurring coagulants have trustworthy and accessible sources.
	• Compactibility with other technologies	Although further research needs to be done on various integrated or hybrid processes using natural coagulants, in general, natural coagulants did not pose significant issues for other treatment approaches.
	• Product stability	Natural coagulants are prone to deterioration, where the active ingredients (protein and polyschaccharides) may deteriorate over time when kept in storage or may change as a result of microbial attack and outside environmental influences during the extraction and purification processes. This is a hurdle for the long-term preservation of natural coagulants because the active components' breakdown will negatively impact their integrity (performance) and create logistical (shipping and storage) issues.

(Continued)

TABLE 3.3 (CONTINUED)
The Sustainability Facets of the Natural Coagulants

S/n	Criteria	Remarks
	• Safety	Due to their natural origin, natural coagulants are typically regarded as harmless and secure. Uncertainty persists regarding the toxicological analysis of natural coagulants and their broader impact on living things. Given that it requires attaching chemicals to the coagulants, this is particularly attractive for the modified natural coagulants. To validate the safety of using natural coagulants, a more thorough assessment of the toxicity of modified and unmodified natural coagulants on humans and the environment is required.
2.	**Environmental**	The use of biodegradable natural coagulants that could reduce sludge production while also not adding to its toxicity may be appealing to water utilities. If the organic sludge produced by natural coagulants could be used for other processes, such as wastewater treatment, the construction industry (cement, concrete, and mortar), agricultural practice, and land-based uses, it might provide environmental sustainability. Rarely is the secondary waste produced by the chemical-based extraction, purification, or synthesis of natural coagulants highlighted.
		Natural coagulant residues are regarded as organic compounds that may interact with disinfectants (produced by a subsequent disinfection process) to produce disinfection by-products that are carcinogenic. The use of natural coagulants should therefore be cautious if a downstream disinfection technique is available. Additionally, contaminants in natural coagulants can be reduced and removed by increasing the purity of the active components.
3.	**Economic**	Natural coagulants have historically been seen as being inexpensive since they may be produced from inexpensive biomass. This assertion, however, is false and oversimplified since it disregards other elements (such as processing costs and regional variations) that have an equal impact on the price of the coagulants. Therefore, it has not been adequately demonstrated that natural coagulants are more affordable than synthetic coagulants.
		On another side, as the majority of stated claims were based on lab-scale observation, it has been challenging to translate the advantages of employing natural coagulants, such as the creation of less sludge and without the need for pH adjustment, into economic advantages.
		To support the indirect cost savings brought on by the advantages of employing natural coagulants, more thorough testing in a pilot setting or actual use is needed.
		Utilizing hybrid coagulants, which lower the cost of consumption, is another option to increase the economic benefits of employing natural coagulants. As an alternative, investigating the versatility of natural coagulants may also result in a reduction in overall costs because numerous therapy functions can be carried out with a single substance in a single treatment unit.
		Because there hasn't been a thorough analysis of the total expenses of the coagulation process, from extraction to the effects on other treatment units, the economic side of natural coagulants cannot be concluded.

(Continued)

TABLE 3.3 (CONTINUED)
The Sustainability Facets of the Natural Coagulants

S/n	Criteria	Remarks
4.	**Social**	
	• Industrial acceptance	The alternative process using natural coagulants can, in comparison to the currently used established process, give equivalent (or better) treatment performance at comparable (or lower) cost (coagulation with chemical coagulants). One of the main reasons the industry is hesitant to use natural coagulants is the absence of pilot scale or actual application of natural coagulants in water and wastewater treatment operations. The absence of regulatory guidelines and approval on the use of natural coagulants, particularly in the process of treating potable water, is another barrier that prevents the industry from accepting them.
	• Public health improvements	People without access to clean water, particularly those who live in rural areas or regions with limited resources for adequate clean water supply, may benefit from using natural coagulants that can be acquired locally and extracted using a straightforward technique. Drinking water that has been treated will enhance their health, which will elevate their living conditions.

Source: Lun and Wahab (2020).

By creating items that can be completely recycled or used again, like those found in many natural adsorbents, cradle-to-cradle design will put a stop to the production of goods that are used once and then thrown away (Oladoja, 2015). The procedure is environmentally benign, and this will boost its economic viability. Another objective is the creative development of alternative technologies that have been shown to be harmful to human health and the environment. The usage of traditional coagulants, such as alum, has been linked to adverse effects on human health, the production and disposal of huge amounts of sludge, and subsequent environmental effects. Additionally, green technology alters the patterns of production and consumption to reduce sources of waste and pollution.

When used to describe environmental technology, the term "green technology" also refers to "clean technology". To monitor, model, and conserve the natural environment and resources, as well as to lessen the negative effects of human involvement, it applies one or more of environmental science, green chemistry, environmental monitoring, and electronic devices. Considering some of the characteristics of biobased coagulants, such as their high biodegradability, non-toxicity, and non-corrosive nature, using them for water and wastewater treatment would assist to realize these aims. Sustainable development is the foundation of green environmental technology.

CONCLUSION

This chapter discussed how biocoagulants and bioadsorbents (BCAs) remediate industrial wastewater and how they help create a greener environment. This involved researching natural precursors and how they were made into BCAs. The use of different BCAs in the treatment of wastewater containing dyes, pharmaceuticals, and other developing pollutants, as well as other industrial effluent, was examined. These natural BCAs' coagulation and adsorption mechanisms, as well as their contribution to a greener environment and attainment of the sustainable development goal, were reviewed. The study showed how effective BCAs are in treating different industrial effluents and how it enhances our ability to achieve cleaner environments.

REFERENCES

Alazaiza, M. Y. D., Albahnasawi, A., Ali, G. A. M., Bashir, M. J. K., Nassani, D. E., Al Maskari, T., … Abujazar, M. S. S. (2022). Application of natural coagulants for pharmaceutical removal from water and wastewater: A review. *Water, 14*(2), 140.

Dalvand, A., Gholibegloo, E., Ganjali, M. R., Golchinpoor, N., Khazaei, M., Kamani, H., … Mahvi, A. H. (2016). Comparison of Moringa stenopetala seed extract as a clean coagulant with Alum and *Moringa stenopetala*-Alum hybrid coagulant to remove direct dye from textile wastewater. *Environmental Science and Pollution Research, 23*(16), 16396–16405. https://doi.org/10.1007/s11356-016-6708-z

Ezemagu, I. G., Ejimofor, M. I., & Menkiti, M. C. (2020). Turbidimetric study for the decontamination of paint effluent (PE) using mucuna seed coagulant (MSC): Statistical design and coag-flocculation modelling Turbidimetric study for the decontamination of paint effluent (PE) using mucuna seed coagulant. *Environmental Advances, 2*(December), 100023. https://doi.org/10.1016/j.envadv.2020.100023

Hamidi, D., Besharati Fard, M., Yetilmezsoy, K., Alavi, J., & Zarei, H. (2021). Application of *Orchis mascula* tuber starch as a natural coagulant for oily-saline wastewater treatment: Modeling and optimization by multivariate adaptive regression splines method and response surface methodology. *Journal of Environmental Chemical Engineering, 9*(1), 104745. https://doi.org/10.1016/j.jece.2020.104745

Jamshidi, A., Rezaei, S., Hassani, G., Firoozi, Z., Ghaffari, H. R., & Sadeghi, H. (2020). Coagulating potential of Iranian oak (*Quercus brantii*) extract as a natural coagulant in turbidity removal from water. *Journal of Environmental Health Science and Engineering, 18*(1), 163–175. https://doi.org/10.1007/s40201-020-00449-0

Kristianto, H., Rahman, H., Prasetyo, S., & Sugih, A. K. (2019). Removal of Congo red aqueous solution using *Leucaena leucocephala* seed's extract as natural coagulant. *Applied Water Science, 9*(4), 1–7. https://doi.org/10.1007/s13201-019-0972-2

Kumar, V., Saidulu, D., Majumder, A., & Srivastava, A. (2021). Emerging contaminants in wastewater: A critical review on occurrence, existing legislations, risk assessment, and sustainable treatment alternatives. *Journal of Environmental Chemical Engineering, 9*(5), 105966. https://doi.org/10.1016/j.jece.2023.105966

Lopes, E. C., Santos, S. C. R., Pintor, A. M. A., Boaventura, R. A. R., & Botelho, C. M. S. (2019). Evaluation of a tannin-based coagulant on the decolorization of synthetic effluents. *Journal of Environmental Chemical Engineering, 7*(3), 103125. https://doi.org/10.1016/j.jece.2019.103125

Lun, W., & Wahab, A. (2020). State of the art and sustainability of natural coagulants in water and wastewater treatment. *Journal of Cleaner Production, 262*, 121267. https://doi.org/10.1016/j.jclepro.2020.121267

Menkiti, M., Ezemagu, I., & Okolo, B. (2016). Perikinetics and sludge study for the decontamination of petroleum produced water (PW) using novel mucuna seed extract collision factor for Brownian transport. *Petroleum Science*, *13*(2), 328–339. https://doi.org/10.1007/s12182-016-0082-9

Mohd Asharuddin, S., Othman, N., Altowayti, W. A. H., Abu Bakar, N., & Hassan, A. (2021). Recent advancement in starch modification and its application as water treatment agent. *Environmental Technology and Innovation*, *23*, 101637. https://doi.org/10.1016/j.eti.2023.101637

Muniz, G. L., Borges, A. C., da Silva, T. C. F., Batista, R. O., & de Castro, S. R. (2022). Chemically enhanced primary treatment of dairy wastewater using chitosan obtained from shrimp wastes: Optimization using a Doehlert matrix design. *Environmental Technology (United Kingdom)*, *43*(2), 237–254. https://doi.org/10.1080/09593330.2020.1783372

Nath, A., Mishra, A., & Prakash, P. (2020). Materials today: Proceedings of the a review natural polymeric coagulants in wastewater treatment. *Materials Today: Proceedings*. https://doi.org/10.1016/j.matpr.2020.03.551

Nicholas, H., Hoong, J., & Ismail, N. (2018). Removal of dye in wastewater by adsorption-coagulation combined system with *Hibiscus sabdariffa* as the coagulant. *MATEC Web of Conferences*, *152*, 01008.

Nnaji, P. C., 2019. Simulated Dye and Textile Wastewater Treatment Using Mucuna Sloanei and Luffa Cylindrica Seeds as Natural Coagulants. Ph.D. Thesis, Chukwuemeka Odumegwu Ojukwu University, Uli, Anambra State, Nigeria

Nnaji, N. J. N., Ani, J. U., Aneke, L. E., Onukwuli, O. D., Okoro, U. C., & Ume, J. I. (2014). Modelling the coag-flocculation kinetics of cashew nut testa tannins in an industrial effluent. *Journal of Industrial and Engineering Chemistry*, *20*(4), 1930–1935. https://doi.org/10.1016/j.jiec.2013.09.013

Nnaji, P., Anadebe, C., & Onukwuli, O. D. (2020a). Application of experimental design methodology to optimize dye removal by *Mucuna sloanei* induced coagulation of dye-based wastewater. *Desalination and Water Treatment*, *198*, 396–406. https://doi.org/10.5004/dwt.2020.26017

Nnaji, P. C., Okolo, B. I., & Onukwuli, O. D. (2020b). *Luffa cylindrica* seed: Biomass for wastewater treament, sludge generation study at optimum conditions. *Chemical Industry and Chemical Engineering Quarterly*, *26*(4), 349–358. https://doi.org/10.2298/CICEQ190623012N

Nnaji, P. C., Anadebe, V. C., Onukwuli, O. D., Okoye, C. C., & Ude, C. J. (2021). Multifactor optimization for treatment of textile wastewater using complex salt–*Luffa cylindrica* seed extract (CS-LCSE) as coagulant: response surface methodology (RSM) and artificial intelligence algorithm (ANN–ANFIS). *Chemical Papers,* 0123456789. https://doi.org/10.1007/s11696-021-01971-7

Nonfodji, O. M., Fatombi, J. K., Ahoyo, T. A., & Osseni, S. A. (2020). Performance of *Moringa oleifera* seeds protein and *Moringa oleifera* seeds protein-polyaluminum chloride composite coagulant in removing organic matter and antibiotic resistant bacteria from hospital wastewater. *Journal of Water Process Engineering*, *33*(December 2019). https://doi.org/10.1016/j.jwpe.2019.101103

Okolo, B. I., Nnaji, P. C., & Onukwuli, O. D. (2016). Nephelometric approach to study coagulation-flocculation of brewery effluent medium using Detarium microcarpum seed powder by response surface methodology. *Journal of Environmental Chemical Engineering*, *4*(1). https://doi.org/10.1016/j.jece.2015.12.037

Oladoja, N. A. (2015). Headway on natural polymeric coagulants in water and wastewater treatment operations. *Journal of Water Process Engineering*, *6*, 174–192. https://doi.org/10.1016/j.jwpe.2015.04.004

Onukwuli, O. D., Nnaji, P. C., Menkiti, M. C., Anadebe, V. C., Oke, E. O., Ude, C. N., … Okafor, N. A. (2021). Dual-purpose optimization of dye-polluted wastewater decontamination using bio-coagulants from multiple processing techniques via neural intelligence algorithm and response surface methodology. *Journal of the Taiwan Institute of Chemical Engineers*, *125*, 372–386. https://doi.org/10.1016/j.jtice.2023.06.030

Owodunni, A. A., & Ismail, S. (2021). Revolutionary technique for sustainable plant-based green coagulants in industrial wastewater treatment – A review. *Journal of Water Process Engineering*, *42*(2), 102096. https://doi.org/10.1016/j.jwpe.2023.102096

Saleh, I. A., Zouari, N., & Al-Ghouti, M. A. (2020). Removal of pesticides from water and wastewater: Chemical, physical and biological treatment approaches. *Environmental Technology and Innovation*, *19*, 101026. https://doi.org/10.1016/j.eti.2020.101026

Santos, A. F. S., Matos, M., Sousa, Â., Costa, C., Teixeira, J. A., Paiva, P. M. G., … Costa, C. (2015). Removal of tetracycline from contaminated water by *Moringa oleifera* seed preparations, *3330*(September), 0–8. https://doi.org/10.1080/09593330.2015.1080309

Sibartie, S., & Ismail, N. (2018). Potential of *Hibiscus sabdariffa* and *Jatropha curcas* as natural coagulants in the treatment of pharmaceutical wastewater. *MATEC Web of Conferences*, *152*. https://doi.org/10.1051/matecconf/201815201009

Yadav, S., Yadav, A., Bagotia, N., Sharma, A. K., & Kumar, S. (2021). Adsorptive potential of modified plant-based adsorbents for sequestration of dyes and heavy metals from wastewater – A review. *Journal of Water Process Engineering*, *42*(March), 102148. https://doi.org/10.1016/j.jwpe.2023.102148

4 Technological Applications for Green and Sustainable Production and Elimination of Environmental Pollution

A. Gürses, K. Güneş, E. Şahin, and M. Açıkyıldız

CONTENTS

DOI: 10.1201/9781003301769-4

4.1 INTRODUCTION

The concept of "green chemistry", which began to be recognized by the scientific world in the early 1990s, was soon adopted as a new, popular, and promising chemistry approach against the common industrial practice of first polluting and then cleaning. The concept is most succinctly defined as the design of chemical products and processes that reduce or eliminate the use or production of hazardous substances (Centi & Perathoner 2003).

Green chemistry leads to the search for new processes and technologies, using harmless or safer than currently used ones as starting materials, which aim to prevent or reduce environmental pollution in a short time by reducing the volumes and toxicities of chemical wastes. Ensuring the development and continuity of "Green Chemistry" in order to create a safer and more sustainable natural environment in the future has now emerged as the most acceptable and rational way. Attractive applications such as heterogeneous catalysis, biocatalysis, solvent-free synthesis, microwave, and ultrasound use are also some typical examples in this context (Mukherjee 2021, Yadav 2006).

Twelve basic principles have been identified for green chemistry, which has become a means of promoting sustainable progress and development at the laboratory and industrial scale. These principles are the primary guidelines for following this approach and form the framework for the measures to be taken to make chemical products and processes more harmless to the environment (Anastas 1998; Poliakoff et al. 2002; Tobiszewski et al. 2015) (Figure 4.1).

1. **Waste prevention:** It is the primary principle of green chemistry, and it refers to designing chemical syntheses to prevent waste without leaving waste to be processed It is correct or cleaned. In this context, it is essential to minimize the amount of solvents, catalysts, and auxiliaries, as well as reactants that cannot become a part of the target product. Regarding waste prevention, it should be kept in mind that the solvents used in the separation and purification steps are also the main sources of waste (Dube & Salehpour 2014; Ardila-Fierro & Hernández 2021).

1. Waste prevention
2. Maximizing atom economy
3. Designing less dangerous chemical syntheses
4. Designing safer chemicals and products
5. Using safer reaction conditions and solvents
6. Increase energy efficiency

7. Using renewable raw materials
8. Avoiding chemical derivatives
9. Using catalysts instead of stoichiometric reagents
10. Designing chemical products that can be degraded after use
11. Real-time analysis to avoid pollution
12. Minimizing the potential for accidents

Principles of Green Chemistry

FIGURE 4.1   Twelve basic principles of green chemistry.

2. **Maximizing atom economy:** Good atom economy means that most of the atoms of the reactants are incorporated into the desired products and only small amounts of unwanted by-products are formed. Already, in practice, the success of a reaction is judged by the percent yield of the product formed. However, the nature and quantity of by-products produced by the reaction are often ignored. This has given rise to the concept of atomic economy, which takes into account the efficiency of incorporating reagents into the final product, thereby reducing the depletion of raw materials and the generation of waste (Trost 1991; Adeleye et al. 2021; 2000).

3. **Designing less dangerous chemical syntheses:** It refers to the design of syntheses to use and produce substances that have little or no toxicity to humans and the environment. Chemical synthesis reactions, which usually take place in multiple steps, use reagents, most of which are toxic, and although the product does not contain these toxic substances directly, there is a risk of containing them as residues, so the processes need to be revised in this context. Less hazardous chemical synthesis relies, whenever possible, on the creation of synthetic methods for the use and production of substances that are little or no toxic to human health and the environment. In this context, for example, the use of alternative biological enzymes instead of harmful chemicals can make many industrial processes cleaner and cheaper (Tundo et al. 2000; Ivanković et al. 2017).

4. **Designing safer chemicals and products:** Designing chemical products that are fully effective but have little or no toxicity. Designing safer chemical products and processes by reducing hazards and toxicity and preventing accidents also means safer work and production.

However, for production management, minimizing toxicity and maintaining efficacy can be significant challenges for the design of safer products and processes. For this, designing safer chemicals requires knowledge of chemical, toxic, and physical properties and various risks (Schulte et al. 2013; Chen et al. 2020). In this context, different levels of hazards have been clearly defined for various chemical production areas and some more advantageous methods for improving the design of safer chemicals have been proposed. These can be listed as more advanced technologies for hazard treatment at the molecular level, toxicity tests to be developed to reduce hazards, with financial benefits of more effective elimination of hazards, and lower environmental impacts (Anastas & Warner 2005). However, it is clear that safer design cannot be achieved without sufficient chemical structure information and the necessity of a real-time monitoring system for process analysis to determine the hazards or toxicity of chemicals. Such an analysis can instantly detect changes in parameters such as pH, gas flow rate, pressure, and temperature before the process becomes unstable, and can definitively prevent the possible destruction of hazardous chemicals.

5. **Using safer reaction conditions and solvents:** It means that the use of solvents, purification or separating agents or other auxiliary chemicals is restricted, and if the use of these chemicals is necessary, the use of safer ones is more appropriate. Solvents and other separating agents adversely affect the general character of the chemical reaction in terms of environmental friendliness or green chemistry. Traditionally, most chemical reactions only take place in the presence of a solvent because the reactants can mix and react efficiently in a homogeneous solution, which speeds up the mixing and enables the reactant molecules to converge quickly and continuously, as well as the uniform heating or cooling of the mixture in a solution can be accomplished more easily. Except for hydrolysis or solvolysis reactions, solvents are normally not incorporated into the product and can be recovered unchanged after the reaction is finished.

It is known that in a standard batch chemical process, solvents make up 50–80% of the total mass, and the solvents used are also responsible for about 75% of the cumulative life cycle environmental impacts (del Pilar Sánchez-Camargo et al. 2019; Bhandari & Raj 2017).

6. **Increase energy efficiency**: Run chemical reactions at room temperature and pressure whenever possible.

Increasing concerns over the depletion of petroleum feedstocks and the growth in energy consumption have pushed the development of additional energy-efficient procedures and the search for renewable energies; non-depleting resources in a certain time frame (Anastas & Eghbali 2010; Laughton 1990). The energy requirements of chemical procedures should be recognized for their economic and environmental impacts. In 1973, the oil crisis has initiated the development of several processes in which energy savings are considered, to exploit every kilojoule of energy in the production process (Ivanković et al. 2017). Unutilized energy may also be considered a waste. The design of chemical reactions that do not involve intensive energy use is highly desirable. Dropping the energy barrier of a chemical reaction or choosing suitable reactants so that the transformation may proceed at room temperature is one example of what chemists can do in order to reduce energetic requirements, with all the direct and indirect benefits accompanying it (Abdussalam-Mohammed et al. 2020). Rising the energy efficiency of a chemical system is purely one part of the solution. Alternative energies are wanted. Several of those renewable energies have been recognized in biofuel production, including solar power (thermal and photovoltaic), wind power, geothermal energy, hydropower, and hydrogen fuel cells. Green chemists have a significant role to play in this new challenge as it is known that they can design both energy-efficient transformations and chemical systems that can be used to harvest some of those renewable natural energies (Foster et al. 2009).

7. **Using renewable raw materials:** It means that in production, starting materials or raw materials must be renewable.

   The source of renewable raw materials is usually agricultural products or other process wastes, while the source of non-renewable raw materials is usually fossil sources such as oil, natural gas, and coal. Chemical reactions should be designed to use renewable but technically and economically viable raw materials and feedstocks. For example, benzene, which is used in the commercial synthesis of adipic acid required for the manufacture of nylon, plasticizers, and lubricants, has been replaced to some extent by renewable and nontoxic glucose, and the reaction can be carried out in aqueous medium (Badami 2008).

8. **Avoiding chemical derivatives:** It refers to minimizing or avoiding, if possible, unnecessary derivatizations such as the use of blocking groups, protection/deprotection, and temporary modification of physical/chemical processes. Derivatization means the use of additional energy and reagents, as well as the generation of extra waste in the synthesis. However, many processes can be designed to reduce the use of additional reagents and the resulting waste. Synthesis of a derivative of the compound may also be necessary (Wardencki et al. 2005; Saleh & Koller 2018). It can also include the use of substances that protect or deprotect, and any short-term changes in the physical and chemical process, and the choice of protecting group is the key factor behind the successful execution of a synthetic process. The overall efficiency and duration of the synthetic process are greatly influenced by the choice of protecting group. The derivation of the selected reactive site so that it becomes more sensitive to the reacting species will promote selectivity in the reaction. Minimizing the use of derivatives in a chemical synthesis can

be achieved by avoiding the use of protecting groups that will increase the atomic economy in the reaction (Kottappara et al. 2020).

9. **Using catalysts instead of stoichiometric reagents:** It means minimizing waste by using catalytic reactions. While catalysts, which reduce the activation energy of the reaction and reduce the amount of energy required for conversion, are substances that are effective in small quantities and can accelerate a single reaction many times, stoichiometric reagents are used in high quantities and only react once. Catalysis is more advantageous than stoichiometric reactions in terms of both selectivity and energy minimization. The specificity of a catalyst is associated with directing the reaction to the desired product, reducing the amount of unwanted by-products, and thus reducing the waste to be produced (Anastas et al. 2000).

10. **Designing chemical products that can be degraded after use:** It refers to designing chemical products to be separated into harmless substances so that they do not accumulate in the environment after use (Logar 2011). In addition to the fact that materials and products are obtained from renewable resources, they should not be permanent in the environment, but unfortunately many products used in daily life are very permanent. This is one of the most difficult principles to apply, although there is solid empirical evidence of which types of substances degrade easily in the environment. Products must be stable long enough to be ready before and during use, and degradation may occur depending on water, light, and oxygen levels and the presence of various microorganisms or other environmental factors (Shonnard et al. 2012). Plastic waste does not degrade easily and pharmaceuticals such as antibiotics accumulate in water bodies. The principle is based on the design of products to perform specified functions and requires the inclusion of cleaning with nontoxic and biodegradable chemicals in the design. Cleaning agents recommended for use in this context are more effective alternatives to traditional cleaning products, with less negative impact on the user and the environment. For example, using water with suitable additives as a solvent to remove grease and oil would be a safer and more convenient alternative to the volatile hydrocarbon solvents typically used (Dua et al. 2012).

11. **Real-time analysis to avoid pollution:** In other words, the principle emphasizes the importance of reaction monitoring, especially in the control of chemical hazards and process safety (Ahuja 2013; Jiang et al., 2021). Moreover, real-time, in-process monitoring can also be used to prevent waste, increase synthetic efficiency, aid catalyst design, and support the use of unconventional or complex techniques such as solvent-free chemistry or biochemical processing (Sans & Cronin 2016). Hazard prevention can be realized by careful control of chemical reactions during production. For this, through a better understanding of reaction mechanisms and intermediates, and the study of interactions between process variables, it may become possible to avoid problematic

situations such as overdosing of reagents, overheating or sub cooling, loss of selectivity, or decomposition of desired products. In most cases, ignoring those leads to chemical waste and pollution, or the formation of toxic substances (Erythropel et al. 2018).

12. **Minimizing the potential for accidents:** It means that solid, liquid, or gaseous chemicals must be designed to minimize the possibility of chemical accidents, including explosions, fires, and uncontrolled releases into the environment. Since this principle is directly based on accident prevention, it can be considered the most fundamental safety-related chemistry principle. Its basic condition is the reduction of the use of substances that can cause adverse effects such as explosion, fire, and harmful vapor formation in chemical processes and effective control of process conditions. Today, the increasing use of supercritical CO_2, which is nontoxic or non-explosive and environmentally friendly, as an alternative to organic solvents is a striking example of the realization of this principle. Safety can be defined as the control of known hazards taking into account an acceptable level of risk. The further level of safety control is specialist engineering controls that refer to the application of physical process modifications, such as reducing contact with hazardous chemicals, isolating the process, using wet methods to reduce dust generation, and aerating (Ivankovic et al. 2017).

Green chemistry applies to all industrial sectors such as automotive, electronics, energy, and space, and its most important aspect is the design philosophy or understanding of its principles, also known as design rules, that enable manufacturers to achieve sustainability. Green chemistry principles, which also provide financial benefits to manufacturers such as less energy and cost, less use of inputs, a safer end product, and a better brand image, are vital in terms of environmental negative effects and increasing the safety of chemical products and processes (Zeng & Li, 2021; Whiteker, 2019; Omodara et al. 2023; Sharma et al. 2008). Since it is aimed to reduce waste at every stage, it can produce more sustainable outputs, but less energy consumption can lead to more investment expenditure. Therefore, the industrial expectation is to reduce waste and thus disposal costs, reduce processing costs due to less energy consumption, minimize the formation of hazardous by-products, and use less amount of raw materials due to increased efficiency (Raj et al. 2022).

Sustainability, eco-efficiency, and green chemistry are key concepts guiding the development of next-generation products and processes. Since the future of the world largely depends on the chemical processes taken into account and used, chemistry can be decisive in terms of understanding and protecting the environment correctly (Sanghi 2000; Anastas & Warner 1998). For this reason, chemistry, and especially green chemistry, which is included in everything around people, is considered the most important part of studies aimed at protecting human health and the environment. Green chemistry, which is a science-based, regulatory, and economic-oriented approach to achieving environmental protection and sustainable development goals (SDGs), can be environmentally friendly or green in practice if it meets at least some, if not all, of its basic principles (Wilkinson 1997; Faber 2011; Sharma et al. 2008).

This chapter is devoted to the synthesis of environmentally friendly chemical products and various sustainable removal processes, as well as the introduction of green sustainable methods for the classification and identification and subsequent removal operation of various metal ions, organic and inorganic industrial pollutants.

4.2 ENVIRONMENTAL POLLUTANTS AND GREEN TREATMENT TECHNIQUES

4.2.1 Inorganic Pollutants

The release of heavy metals and metal compounds into water, soil, and air is one of the most worrying environmental pollutants (Briffa et al. 2020). Intensive industrial, metallurgical, and mining activities inevitably lead to the emergence of various elemental metals and metal compounds, which have adverse effects on the environment and living things (Li et al. 2019). Heavy metal ions such as iron (Fe^{2+}), nickel (Ni^{2+}), copper (Cu^{2+}), zinc (Zn^{2+}), silver (Ag^+), cadmium (Cd^{2+}), lead (Pb^{2+}), and mercury (Hg^{2+}) are especially serious due to their persistence in common environmental conditions. They are categorized as hazardous inorganic pollutants. Heavy metals, characterized by high toxicity, persistence in the environment, and bioaccumulation in plant and animal tissues, have characteristic physicochemical properties (Ali & Khan 2018; Colozza et al. 2021). Various techniques such as precipitation, oxidation, reduction, filtration, ion exchange, reverse osmosis, membrane filtration, adsorption, electrocoagulation, and electroflotation are used to remove heavy metals from aqueous solutions.

Ions such as nitrate, ammonium, and phosphate are descriptors of water quality and good ecological equilibria in water. However, excessive amounts of these components, especially due to excessive use of fertilizers in agricultural applications, can change the optimal equilibrium conditions and cause eutrophication, creating another category of pollutants (Cuartero et al. 2020; Sinha et al. 2017). On the other hand, different forms of these components, especially nitrites and ammonia, become harmful to the aquatic environment (Kim et al. 2019).

4.2.2 Organic Pollutants

Pesticides are chemical substances used to destroy and mitigate or control pests, including herbicides, insecticides, nematicides, molluscicides, piscicides, avisides, rodenticides, bactericides, insect repellents, animal repellents, germicides, fungicides, and lampricides. The most common of these are herbicides, accounting for about 4/5 of all pesticide use. Most pesticides are designed to serve as plant protection products that generally protect plants from weeds, fungi, or insects. Among these chemicals, those that may contain organochlorine pesticides, pyrethroids, organophosphorus, chloronitrophenols, and heavy metal ions can have serious harmful effects on the environment and living things even when present at trace levels (Ma et al. 2020; Reyes-Calderón et al. 2022). Pesticides, which are widely used in a wide range of applications, can resist for a long time in different environments, such as biological materials, wastes, vegetables, and fruits, and in environments such as water, air, and soil they can reach due to their high consumption rates. Generally, they have dangerous side effects and can cause a number of harmful diseases even at low concentrations (Bagheri et al. 2021).

Population growth all over the world and the consequent increase in the consumption of pharmaceuticals including personal care products, hormones, antibiotics, and drugs have led to the availability of these products in high concentrations in the environment. For example, it is estimated that the annual growth rate of the pharmaceutical industry is 6.5% and 10% of the manufactured pharmaceutical products posing an environmental hazard (Nyaga et al. 2020).

These pharmaceutical products and their residues can reach the environment in different ways, such as wastewater from pharmaceutical production facilities, domestic and hospital wastewater containing pharmaceutical compounds, the use of veterinary drugs and feed additives, and improper disposal of unused or expired pharmaceutical products (O'Flynn et al. 2021).

4.2.3 GREEN PRE-CONCENTRATION AND REMOVAL TECHNOLOGIES

Due to their destructive effects and toxicity, advanced technological advances have been made for the analysis of a variety of harmful chemicals, including herbicides and pesticides, phenolic compounds, polycyclic aromatic hydrocarbons, and heavy metals (Petridis et al. 2014; Chatzimitakos et al. 2016; Samadi et al. 2012 ; ;Hashemi et al. 2015; Zhao et al. 2012). The demand for analytical methods for the accurate and rapid determination of harmful compounds in the environment has also increased dramatically in recent years. Despite these recent advances, sample pretreatment steps are still required for accurate quantification of analytes in environmental samples. Therefore, it is extremely important to develop efficient, cost-effective, and simple procedures for the pretreatment, pre-concentration, and separation of trace contaminants in various environmental samples (Hashemi et al. 2017). Under certain conditions, they may require pre-concentration as the level of trace elements can be very low, and performing pre-concentration procedures is very useful for the analysis of environmental samples. The most commonly used methods for this purpose are solvent extraction, cloud point extraction, solid-phase extraction (SPE), ion exchange, solvent sublation, and electrodeposition (Dhankhar & Hooda 2011; Vijayaraghavan & Yun 2008; Liu & Liu 2008; Chojnacka 2010; Dercová et al. 2005; Türker 2007; Lemos et al. 2008). Increasing environmental concerns have increased the need for effective and selective separation techniques for metal ions.

4.2.3.1 Solid-phase extraction (SPE)

Solid-phase extraction is a process that uses a solid phase and a liquid phase to extract desired components from a solution. This process, which has the advantages of being more precise, easy, environmentally friendly, and faster, is recognized as an effective tool for the separation or pre-concentration of various inorganic and organic compounds (de Godoi Pereira & Arruda 2003). Solid-phase extraction (SPE) is known as a widely used green chemistry method for the separation and pre-concentration of metals in analytical chemistry (Wan Ibrahim et al. 2014; Okenicová et al. 2016).

4.2.3.2 Magnetic solid-phase extraction (MSPE)

The classical solid-phase extraction technique has a number of limitations, including retention of sorbent particles in cartridge frits that create backpressure and creation of flow channels in the solid-phase extraction column, which leads to poor retention of target components (Huang et al. 2019a). In addition, in the solid-phase extraction technique, it is necessary to separate the adsorbent from the solution using methods such as filtration, precipitation, and centrifugation, and these processes are time consuming, and there is also the possibility that a part of the adsorbent remains in the solution, thus reducing the performance of the applied method (Ghiasi et al. 2020; Zhang et al. 2020). In contrast, the use of magnetic solid-phase extraction (MPSE), in which the adsorbent is recovered from the sample using an external magnet, not only eliminates these limitations but also reduces the extraction time and improves performance (Zhang et al. 2019).

4.2.3.3 Dispersive solid-phase extraction (dSPE)

This is a popular and versatile pretreatment technique due to its simplicity, low cost, speed, and ease of use. In dispersive solid-phase extraction (dSPE), the adsorbent is added directly to the solution and dispersed homogeneously in order not only to improve the interactions between the adsorbent and the target components, but also to shorten the extraction time. After dispersion of the adsorbent by agitation, ultrasonic or vortex mixing, and subsequent interaction with the target components, the adsorbent is collected by centrifugation, precipitation, or the use of a magnet. One of the processes that could be effective to increase the interactions between the adsorbent and the components is ultrasonic treatment, which can also shorten the extraction times. After separation of the adsorbent, the adsorbed components are desorbed using a small volume of a suitable solvent (Gao et al. 2020; Pan et al. 2020).

4.2.3.4 Solid-phase micro-extraction (SPME)

Solid-phase micro-extraction (SPME), as an environmentally friendly technique, was first developed by Pawliszyn and Arthur in 1990 and received great attention. The main feature of this technique is that it can combine sampling, extraction, and sample introduction in a single step. Covalent organic frameworks (COFs) are used as adsorbents in SPME, particularly for the extraction of pesticides (Feng et al. 2020; Bagheri et al. 2021).

4.2.3.5 Liquid–liquid extraction (LLE)

Liquid–liquid extraction (LLE), one of the most common purification techniques used in the preparation of pharmaceuticals and other chemicals, on both laboratory and industrial scales, relies on the difference in solubility of species in a mixture, typically in two immiscible liquids, one reaction mixture and the other extraction solvent. Liquid–liquid extraction often emerges as the most economical technique when a moderately concentrated species needs to be separated from a feed and a suitable strategy is in place to recover the extraction solvent. It may also play an important role in establishing greener chemical processes, in which crude oil-based solvents are replaced by renewable raw materials (Lebl et al. 2019). However, the disadvantage of the technique is that the conventional extraction solvents typically used in industry are organic solvents such as sulfolane tetraethylene glycol and N-methylpyrrolidone (NMP), which require additional investments and additional energy consumption to recover and are considered toxic, volatile, and/or flammable. Consequently, their replacement by more environmentally friendly alternatives has received more attention in recent years. In this context, ionic liquids (ILs), which stand out for their extraordinary combination of properties such as high dissolving capacity, negligible vapor pressure, and low flammability for a wide variety of inorganic and organic materials, have been proposed as very powerful and promising alternatives to conventional organic extractors (Gouveia et al. 2016).

Solvents, which are widely used in many industrial processes, are estimated to account for almost 60% of all industrial emissions and 30% of all volatile organic compound emissions worldwide. Therefore, solvents are of great interest for green chemistry (Pollet et al. 2014; Earle & Seddon 2000; Pena-Pereira et al. 2015; Clark et al. 2015). Large volumes of solvents typically used in a reaction, especially in the purification step or in a formulation, are not directly responsible for the composition of a reaction product, nor are they the active ingredient of a formulation. Therefore, the use of toxic, flammable, or environmentally harmful solvents seems unnecessary because these properties have no effect on the function or progress of the system in which the solvent is used (Abou-Shehada et al. 2016; Constable et al. 2007; Byrne et al. 2016).

Studies of sustainable solvents and increasing awareness of the effects of solvents on pollution, energy use and air quality, and climate change have led to increased scientific and industrial research. Solvent losses account for a large proportion of organic contamination and a large proportion of solvent removal and process energy consumption.

To overcome these problems, a number of greener or more sustainable solvents have been developed in recent years. While how a substance is distributed is as important to sustainability as what it is made of, the focus is on the environmental credentials of the solvent itself (Clarke et al. 2018). It is recognized that many commonly used solvents are of environmental concern due to three main reasons: origin and synthesis of the solvent itself, disruptions and difficulties in use, and disposal. It is argued that the solution to these concerns depends on a solvent or solvent class being green in nature. Solvents and solvent classes designated as green solvents include water, supercritical fluids, gas-expanding fluids, ionic fluids, liquid polymers, and biomass-derived solvents (Akiya & Savage 2002; Simon & Li 2012; Lindström 2002; Dallinger & Kappe 2007; (Beckman) 2004; Rayner 2007; Han & Poliakoff 2012; Boyère et al. 2014; Jessop & Subramaniam 2007; Pârvulescu & Hardacre 2007; van Rantwijk & Sheldon 2007; Plechkova & Seddon 2008; Hallett & Welton 2011; Naughton & Drago 1995; Chandrasekhar et al. 2002; Feu et al. 2014; Mathers et al. 2006; Spear et al. 2007; Horváth 2008). This is based on the notion that replacing a non-green solvent in a process with a solvent qualified as green will necessarily improve environmental performance. However, which solvents are more environmentally friendly is also open to debate (Clark & Tavener 2007). Although water exhibits similar properties, ionic liquids in particular have often been criticized for their complex synthesis and toxicities (Jessop 2011; Blackmond et al. 2007; Welton 2015).

4.2.3.6   Ionic liquids (ILs)

The terms room-temperature ionic liquid, non-aqueous ionic liquid, molten salt, liquid organic salt, and fused salt are all used to describe salts in the liquid phase, and ionic liquids are composed of anions and cations (Welton 1999). In contrast to molecular solvents consisting of neutral species such as benzene, methanol, chloroform, and water, common molten salts exhibit a high melting point that greatly limits their use as solvents in most applications. For example, the melting temperatures for sodium chloride and lithium chloride are 801°C and 614°C, respectively. Room-temperature ionic liquids continue the liquid behavior at or below room temperature. The specified upper temperature limit for ionic liquid classification is 100°C and ion systems with higher melting points are normally defined as molten salts (Vishwakarma 2014).

In recent decades, ionic liquids (ILs), consisting entirely of ions and organic salts with melting points below 100°C, have received increasing attention in scientific and industrial fields. Their properties that make them both technologically and ecologically attractive are negligible volatility; non-flammability; and thermal, chemical, and electrochemical stability.

It is possible to design a solvent for a particular use, as they are composed of ionic species that can be exchanged and upgraded and usually consist of bulky organic cations and organic or inorganic anions (Earle & Seddon 2000; van Rantwijk & Sheldon 2007; Welton 2015). Table 4.1 provides the common cations and anions involved in the formation of ionic liquids.

Ionic liquids, which attract great attention in the field of electrochemistry due to their interesting properties such as non-volatility and high ionic conductivity, are considered as suitable electrolytes that can be used in various energy storage devices such as batteries, solar cells, fuel cells, thermoelectrochemical cells, and supercapacitors. Ionic liquids are also used as additives to improve the quality and performance of various substances such as lubricants, dyes, shampoos, softeners, adhesives, and detergents (Kaur et al. 2022). Figure 4.2 schematically shows the major application areas of ionic liquids.

4.2.3.7   Supercritical and subcritical fluids

Gases that are compressed above their critical temperatures, whose densities approach that of their liquid states, behave as supercritical fluids, and many substances have critical temperatures close to room temperature (Poliakoff et al. 1999).

TABLE 4.1
Major Common Cations and Anions Involved in the Formation of Ionic Liquids

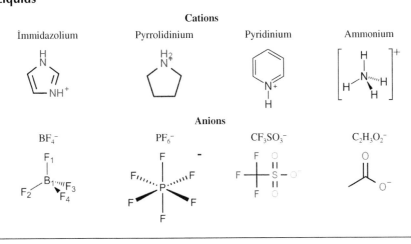

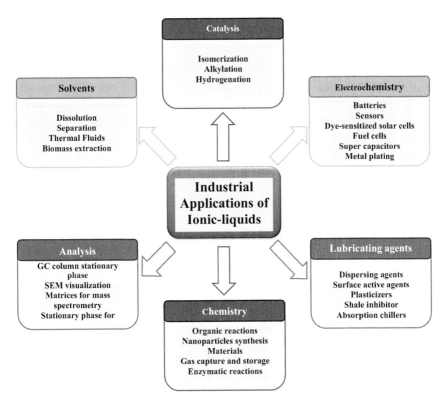

FIGURE 4.2   Schematic representation of the main application areas of ionic liquids.

Supercritical CO$_2$

Supercritical fluids such as carbon dioxide and water are environmentally friendly substances. CO_2, which is gaseous at room temperature and atmospheric pressure, is a small, linear molecule that can diffuse faster as a supercritical fluid than conventional liquid solvents, thus offering a very high extraction power (Herrero et al. 2010). It is often used as a green alternative to toxic Freon's and certain organic solvents. While supercritical CO_2 has good solvent properties for the extraction of nonpolar hydrocarbon compounds, its high electrical moment also allows it to dissolve some moderately polar compounds such as alcohols, esters, aldehydes, and ketones (Lang & Wai 2001). Supercritical carbon dioxide dissolves polar or low polarity compounds, exhibits high solvent power for low molar mass compounds, has high affinity for organic compounds with medium molar mass and oxygen, and also has the ability to separate compounds that are less volatile, higher molar mass, and more polar (Brunner 2005).

Extraction procedures involving supercritical CO_2 provide clean technologies with no secondary products that can pollute the environment (Zeković et al. 2000). Carbon dioxide is generally accepted as a safe solvent, and therefore products extracted with CO_2 in foods do not pose any problems for human health (Rutkowska & Stolyhwo 2009). Also, given the growing concerns about global warming and greenhouse gas emissions, it becomes even more important that potential large-scale supercritical CO_2 processing plants capture the carbon dioxide produced by other industries and create new use fields, rather than releasing it into the atmosphere (Cvjetko Bubalo et al. 2015).

4.2.3.8 Adsorption

Water pollution from heavy metal ions, which have high toxicity, carcinogenicity, and non-degradation properties, has caused a serious environmental concern in recent years (Schwarzenbach et al. 2006). Toxic heavy metals such as lead (Pb), mercury (Hg), cadmium (Cd), chromium (Cr), arsenic (As), zinc (Zn), copper (Cu), and nickel (Ni) are extremely harmful to aquatic life and also cause various diseases and disorders by entering the human body through the food chain (Islam et al. 2015). Accordingly, it is clear that wastewater containing toxic heavy metal ions must be removed from the wastewater effluent before being discharged into the environment. For this, many heavy metal removal methods such as chemical precipitation, adsorption, membrane filtration, electrochemical treatment, coagulation, and flocculation have been extensively investigated (Ge & Li 2018). However, it is suggested that the most efficient and environmentally friendly way to remove metal ions is the adsorption technique, as it has repeatability, low cost, and very high efficiency, as well as simple and less energy requirement (Zhang et al. 2012; Chowdhury et al. 2019; Kebede et al. 2018). Adsorption, which is defined as an increase in the amount of a certain component in the gas or liquid phases at an interface where the solid phase is included, or as a relative increase in the amount of liquid at the interface where the two fluid phases are located, is a widely used method to remove various components from the gas or liquid phases and change the interfacial properties (Dąbrowski 2001). The energetically and morphologically heterogeneous surfaces of the solid phase, which is generally adsorbent, are characterized by active sites interacting with the adsor-bate, which can be atoms, molecules, and ions in the adjacent phase. The process, which depends on many factors that can change the distribution of the adsorbate at the interface between the contacting phases, proceeds with the increase in the amount of adsorbate at the interface as the adsorbate moves from one phase to the other (Weber et al. 1991).

On the other hand, by changing some parameters such as pressure or concen-tration, temperature, and pH, the uptake efficiency of the components adsorbed at the interface can change, and it is possible to transfer or desorb the adsorbed species from the interface to the liquid phase. Biosorbents such as activated carbon and biomass, with their availability, economic feasibility, regenerative capabilities, and environmental friendliness, and green origins, clays, various minerals, polymers, and nanocomposites with high adsorption uptake capacity have been successfully used to remove hazardous pollutants from wastewater (Ighalo et al. 2022).

4.2.3.9   Activated carbon

Activated carbon, a highly porous, high surface area, and non-selective adsorbent material in line with conventions on environmental sustainability, can be seen as a valuable element that will continue to gain interest and applicability in directly or indirectly achieving a sustainable future (Adeleye et al. 2021). Activated car-bon (AC) is a porous, carbonaceous material with ever-expanding applications in water purification and desalination, wastewater treatment, and air purification due to its unique properties.

Moreover, the unique adsorption properties of activated carbon, which may contain functional groups such as carboxyl, carbonyl, phenol, lactone, and qui-none, which are responsible for adsorbing pollutants, are mainly due to activation processes, precursors, and thermal processes (Heidarinejad et al. 2020).

Activated carbon can be produced from agricultural residues such as maize, biomass, rice, maize stalks, pulp, seeds of fruits such as cherries, apricots and grapes, fruit pulp and hard shells such as peanuts, almonds, coffee, and walnuts. The raw material used in the preparation of activated carbon should be abun-dant, cheap, and safe, and its mineral content and biodegradability during storage should be minimal (Prauchner et al. 2016).

Activated carbon is a favorite adsorbent, which is also used for the extraction or pre-concentration of trace elements. The selectivity of activated carbons toward certain components can be improved and the type and distribution of surface functional groups are changed by chemical modification (AlQadhi & AlSuhaimi 2020). On the other hand, especially chemical activation processes are applied to obtain high surface area and increase porosity of activated carbons (Bedia et al. 2018). The production of bio-based activated carbon, other value-added products, and biofuels produced through the use of lignocellulosic biomass/agricultural waste and other waste streams is undoubtedly more in line with the principles of green chemistry.

This type of production offers advantages in terms of environmental protec-tion as well as environmental friendliness in the context of circular economy and sustainability. The use of bio-derived activated carbon in catalytic systems is also promising and has the potential to play a key role in developing a sustainable future (Adeleye et al. 2021).

4.2.3.10 Biochar

Biochars, which are generally produced from agricultural residues such as sugarcane bagasse, cotton stalks, rice straw, among other organic wastes, are adsorbents that can be successfully used to remove metals and organic pollutants from wastewater (Teng et al. 2020).

In addition to pyrolysis conditions such as temperature and heating rate, different precursor types can be used to produce biochar with different properties in terms of surface area and pore size distribution, which can increase the capacity of this green adsorbent, which can be used especially for wastewater treatment (Macedo et al. 2021).

In fact, since it is often difficult to obtain desired physicochemical properties such as high porosity, high specific surface area, and surface functional groups, many studies have been proposed to increase biochar efficiency, including pretreatment/activation/functionalization steps of biochar before the desired application (Lonappan et al. 2016). Most of these studies have focused on increasing the surface area and functional groups, and thus surface chemistry may change during the activation process (Divyashree & Hegde 2015).

Generally, energy-intensive and costly chemical activation involves mineral acids, which, contrary to the principles of green chemistry, raises toxicological and environmental concerns due to their toxic/hazardous effects (Naghdi et al. 2017). The application of organic acids, which are specified as green and sustainable materials due to their biological origins, can be effective to form specific functional groups such as carboxylic acid (–COOH) groups on the biochar surface. The functionalization process is actually a surface phenomenon, and through functionalization, the type and population of functional groups on the surface of biochar can be changed (Mallakpour & Soltanian 2016; Khan & Iqbal 2016; Lonappan et al. 2020).

In addition, biochar has been shown to be an effective sorbent for the absorption of organic pollutants as well as a wide variety of inorganic pollutants from aqueous solutions (Godwin et al. 2019).

4.2.3.11 Clays and minerals

Various modification methods have been used and very successful results have been obtained for a long time to increase the capacities, especially the surface area and cation exchange capacities of clay minerals, which have relatively low adsorption capacity in raw form. Thermal modification, acidic modification, and various organic and nanoparticle-added modifications are the main methods used (Gu et al. 2019; Otunola & Ololade 2020). The most widely used and most effective clay minerals for heavy metal absorption in soil and water are bentonite, montmorillonite, and attapulgite. These clay minerals are often preferred because of their high specific surface area, high cation exchange capacity (CEC), and high swelling property. Different treatment methods can effectively remove various heavy metals, but clays have also been found to effectively adsorb different types of heavy metals from soil, water, wastewater, and sludge.

4.2.3.12   Carbon-based, metal-based, and polymeric nanoadsorbents

Since polymers containing electron-donating functional groups have excellent adsorption capacity as well as high selectivity toward metal ions, studies on the use of various synthetic polymer-based nanocomposites as adsorbents for heavy metal ions have been increasing recently. However, petrochemical-based synthetic polymers have the potential to cause environmental pollution and have low porosity and poor physical and chemical reactivation ability. In contrast, bio-based polymers have excellent porosity and excellent reactivation to pollutants (Naushad et al. 2019). Carbon-based nanoadsorbents such as carbon nanotubes offer a high and effective adsorption capacity for the removal of organic pollutants, as they have large specific surface area and effective interactions with various organic pollutants in the form of hydrophobic, electrostatic, π–π, hydrogen, and covalent bonding (Yang and Xing 2010). Therefore, carbon nanotubes with surface rich in π-electrons allow π–π interactions with organic molecules with C=C bonds or benzene rings, such as polycyclic aromatic hydrocarbons and polar aromatic compounds. Organic compounds with carboxyl, hydroxyl, or amine functional groups can also form hydrogen bonds with the electron-donating graphitic carbon nanotube surface (Yang et al. 2008). In addition, the adsorption of positively charged organic chemicals, such as some antibiotics, can be improved at appropriate pH due to electrostatic attraction (Ji et al. 2009).

In addition, since oxidized carbon nanotubes with carboxyl, hydroxyl, and phenolic surface functionalities can adsorb metal ions with high efficiency, they can offer high efficiency for heavy metal removal by electrostatic attraction and chemical bonding (Rao et al. 2007).

Therefore, it can be said that surface oxidation can significantly increase the adsorption capacity of carbon nanotubes, and moreover, many studies show that they are better adsorbents than activated carbon for heavy metals such as Cu^{2+}, Pb^{2+}, Cd^{2+}, and Zn^{2+}. Surface modification of oxidized carbon nanotubes with anionic polysaccharides is also an alternative way to improve the adsorption performance of carbon nanotubes for the removal of other cationic pollutants from wastewater (Lu et al. 2006). Metal oxide-based nanoadsorbents such as iron oxide, titanium dioxide, and alumina are also effective and low-cost adsorbents for heavy metals such as arsenic, lead, mercury, copper, cadmium, chromium and nickel, and radionuclides (Sharma et al. 2009). Adsorption mainly relies on a complex interaction between dissolved metal ions and oxygen in metal oxides, and the progression of the interaction involves rapid adsorption of metal ions on the outer surface followed by rate-limiting intra-particle diffusion through the micropore walls (Trivedi & Ax 2000). The same process can offer higher adsorption capacity and faster adsorption kinetics with nanoscale adsorbents due to higher specific surface area, shorter intra-particle diffusion distance, and greater number of active surface sites. When the particle size of nanomagnetite decreased from 300 nm to 11 nm, the arsenic adsorption capacity increased more than 100 times, reminiscent of the nanoscale effect. This effect can be explained by the change of magnetite surface structure to form new empty adsorption sites (Auffan et al. 2009). In addition to the high adsorption capacity, some iron oxide nanoparticles such as nanomaghemite and nanomagnetite can be superparamagnetic. Magnetic particles that become superparamagnetic as they get smaller in size can provide easy separation and thus recovery with low-gradient magnetic field. These magnetic nanoparticles can be used directly as adsorbents or as core material in core–shell nanoparticle structure, where the shell provides the desired function, while the magnetic core performs the magnetic separation (Villaseñor & Ríos 2018).

4.2.3.13   Biosorbents

Biosorption can be defined as the removal of certain components from a solution by physical or chemical interactions with biological materials. The process, in which both living and dead organisms can be used, is a promising biotechnology due to its simplicity, apparent efficiency, and prevalence of biomass and waste bioproducts. Various waste biomaterials, microorganisms, bacteria, fungi, yeast, and algae have been used to remove heavy metal ions from wastewater. Besides the treatment of contaminated wastewater, biosorption offers a sustainable waste management tool (Ibisi & Asoluka 2018; Anastopoulos et al. 2019).

Biosorbents have high metal retention and can reduce the concentration of heavy metal ions in solution to trace amounts. This technology has advantages over conventional technologies such as low cost, high efficiency, minimization of chemical and/or biological sludge, no need for additional nutrients, regenerability, and possibility of metal recovery after adsorption (Mahamadi 2011). Algal, fungal, agricultural wastes and residues and bacterial stains are among the examined biosorbents which have satisfactorily been applied to decontaminate waters and wastewaters polluted with heavy metals (Anastopoulos et al. 2019).

An economical sorbent capable of pre-concentrating low levels of different metal ions with relatively high efficiency is of great importance (Dan et al. 2021).

Studies on the removal of a wide variety of pollutants using various green adsorbents are given in Table 4.2.

4.2.4   Catalysis in Green Chemistry

While the environmental budget is often increasingly larger than the R&D budget in the chemical industries, the industrial objectives of R&D have gradually shifted from the medium to the short term, without corresponding compensation by public investment in long-term research. Process innovation is a long-term investment, especially as it involves a complete redesign of the process, and thus industrial research activities in green chemistry are more focused on finding short-term solutions that involve the substitution of hazardous chemicals and solvents rather than developing new solutions that require much higher investments. In addition, the majority of its environmental budget is devoted to

TABLE 4.2

Studies on the Removal of a Wide Variety of Different Types of Pollutants Using Various Green Adsorbents

Adsorbent		Pollutants	References
Guava (*Psidium guajava*) leaves		Cr^{4+}	Mitra and Das (2019)
Sesame hask		Cu^{2+}	El-Araby et al. (2017)
Tobacco leaves		Pb^{2+}	Yogeshwaran and Priya (2021)
Tectona grandis L.f. leaves		Cu^{2+}	Rathnakumar et al. (2009)
Melaleuca diosmifolia leaves		Cr^{4+}	Kuppusamy et al. (2016)
Rubber (*Hevea brasiliensis*) leaf		Pb^{2+}	Kamal et al. (2010)
Bamboo leaves		Hg^{2+}	Mondal et al. (2013)
Phoenix tree leaves		$Cu^{2+}, Cd^{2+} Zn^{2+}$	Yu et al. (2015)
Black tea leaves		Pb^{2+}	Zuorro and Lavecchia (2010)
Cinnamomum camphora leaves		$Cu^{2+}, Cd^{2+} Ni^{2+}$	Wang and Wang (2018)
Arborvitae leaves	Metal ions	$Pb^{2+} Cu^{2+} Co^{2+}$	Shi et al. (2016)
Globimetula oreophila leaves		Cu^{2+}	Chijioke et al. (2021)
Teak (*Tectona grandis*) leaves		Zn^{2+}	Ajmal et al. (2011)
Activated carbon from palm tree leaves (*Phoenix dactylifera*)		Pb^{2+}	Elhussien et al. (2017)
Banana leaves		Cu^{2+}	Darweesh et al. (2022)
Tea leaves (*Camellia sinensis*)		$Pb^{2+} Fe^{2+} Cd^{2+}$	Yue et al. (2016)
Oil palm bagasse (*Elaeis guineensis*)		Cd^{2+}	Villabona-Ortíz et al. (2021)
Moringa oleifera tree leaf powder		$Pb^{2+} Zn^{2+}$	Jayan et al. (2021)
Posidonia oceanica		Pb^{2+}	Elmorsi et al. (2019)
Aloe vera leaves		Cr^{6+}	Prajapati et al. (2020)
Carbonized *Acacia nilotica*		Zn^{2+}	Telkapalliwar and Shivankar (2018)
Mango peel		$Ni^{2+} Zn^{2+} Cu^{2+}$	Iqbal et al. (2009)
Bael fruit		$Pb^{2+} Cu^{2+}$	Barkade et al. (2022)

(Continued)

(Continued)

TABLE 4.2 (CONTINUED)

Studies on the Removal of a Wide Variety of Different Types of Pollutants Using Various Green Adsorbents

Adsorbent	Pollutants	References
Graphene oxide iron-nanoparticles (GS-GO@FeNPs)	Cr^{6+}	Kabir et al. (2022)
Modified biochar	Cr^{6+}	Ambika et al. (2022)
Magnetic metal-organic framework (Fe_3O_4-ZrMOF)	Pb^{2+} Hg^{2+} Cd^{2+}	Ragheb et al. (2022)
Kaolinite-cellulose/cobalt oxide green nanocomposite (Kao-Cel/Co_3O_4NC)	Pb^{2+} Cd^{2+}	Hussain et al. (2022)
Chitosan/Al_2O_3/Fe_3O_4	Zn^{2+} Cu^{2+} Cd^{2+}	Karimi et al. (2022)
Montmorillonite clay	Cr^{3+}	Essebaai et al. (2022)
Multifunctional hydrogel membrane (PVA-CS-LDHs)	Pb^{2+}	Chen et al. (2023)
Magnetite nanoparticles (Fe_3O_4-NPs)	Hg^{2+}	Gindaba et al. (2023)
Demethylated lignin (DAL)	Cr^{6+}	Li et al. (2023)
Magnetic poly(acrylated-menthol deep eutectic solvent) hydrogel	Multi-class pesticide	Jamshidi et al. (2020)
Organic Pollutants		
Pesticides		
DES magnetic bucky gel/multiwalled carbon nanotube nanocomposite	Organochloride	Yousefi et al. (2017)
Polydopamine-based materials	Organochlorine	Musarurwa and Tavengwa (2021).
Carbonized polydopamine	Organochloride	Huang et al. (2015)
Poly(2-aminoterephthalic acid-*co*-aniline) nanocomposite	Organophosphorus	Nasiri et al. (2022)
Magnetic graphene/polydopamine nanocomposite	Benzoylurea	Huang et al. (2019)b
Magnetic zeolitic imidazolate framework-8@deep eutectic solvent	Pyrethroid	Liu et al. (2019a)
Fe_3O_4/graphene nanocomposite	Triazine	Boruah et al. (2017)
Polymeric cryogels modified with cellulose nanomaterials	Atrazine	Tüysüz et al. (2022)
Amino-functionalized magnetic covalent organic framework composite	Sulfonylurea	Tian et al. (2022)

TABLE 4.2 (CONTINUED)
Studies on the Removal of a Wide Variety of Different Types of Pollutants Using Various Green Adsorbents

Adsorbent	Pollutants	References
Poly(ionic liquids) functionalized magnetic nanoparticles	Pyrethroids	Wang et al. (2020)
Magnetic MWCNT functionalized with proline/propylene glycol	Multi-class pesticide	Zhao et al. (2020)
Deep eutectic solvent micro-functionalized graphene (DES-G)	Pyrethroids	Song et al. (2019)
Carbon nanotubes impregnated with metallic nanoparticles	Glyphosate-based herbicides	Diel et al. (2021)
Graphene oxide (GO) and reduced graphene oxide (rGO)	Methomyl, acetamiprid, azoxystrobin	Shi et al. (2022)
Green-synthesized copper nanoparticles (Cu-NPs)	nonsteroidal anti-inflammatory drugs (NSAIDs) (ibuprofen, naproxen diclofenac)	Husein et al. (2019)
γ-Fe$_2$O$_3$/MWCNT/Ag nanocomposites	Pharmaceutical pollutants	Sulfamethazine	Khalatbary et al. (2022)
Bio-based ceramic/organic xerogel (BCO-xerogel)	Antibiotics, amoxicillin (AMX), tetracycline (TC), cefalexin (CLX), and penicillin G (PEN G)	Arabkhani and Asfaram (2022)
Modified activated carbons from potato peels	Pramipexole dihydrochloride dorzolamide	Kyzas and Deliyanni (2015)
Cross-linked with glutaraldehyde and grafted with sulfonate (CsSLF) and N-(2-carboxybenzyl) groups (CsNCB)	Pramipexole dihydrochloride (PRM)	Kyzas et al. (2013)
Multiwalled carbon nanotubes (MWCNTs)	Ciprofloxacin, diclofenac sodium, losartan, ofloxacin and propranolol, and bisphenol-A	Spaolonzi et al. (2022)

(Continued)

TABLE 4.2 (CONTINUED)
Studies on the Removal of a Wide Variety of Different Types of Pollutants Using Various Green Adsorbents

Adsorbent	Pollutants	References
Iron oxide/biochar (Fe_2O_3/biochar)	Nonsteroidal anti-inflammatory drugs (NSAIDs) (salicylic acid, naproxen, and ketoprofen)	Anfar et al. (2020)
Fe_3O_4/biochar	Ibuprofen and acetylsalicylic acid	Liyanage et al. (2020)
Sulfonated spent coffee waste (SCW-SO_3H)	Antibiotics (tetracyclines (TC))	Ahsan et al. (2018)
Mesoporous activated carbons (Meso-AC), MWCNTs)	Carbamazepine and dorzolamide	Ncibi and Sillanpää (2017)
Silver (Ag) nanoparticles decorated reduced graphene oxide (Ag-RGO)	Naproxen	Mondal et al. (2020)
Genipin-crosslinked chitosan/graphene oxide-SO_3H (GC/MGO-SO_3H) composite	Ibuprofen Tetracycline	Liu et al. (2019b)
Silica-clay nanocomposites	Antibiotics (doxycycline, ciprofloxacin, danofloxacin, and sulfamethoxazole)	Levard et al. (2021)
MgO/chitosan/graphene oxide	Ciprofloxacin and norfloxacin	Nazraz et al. (2019)
Horseradish peroxidase immobilized chitosan cross-link with graphene oxide nanocomposite	Ofloxacin	Suri et al. (2021)
Bio-based ceramic/organic xerogel	Tetracycline, amoxicillin, cefalexin, penicillin G	Arabkhani and Asfaram (2022)
The *Moringa* seeds waste (MSW)	Congo Red Disperse Red 60	khamis Soliman et al. (2019)
Ipomoea aquatica root	Basic Yellow 2	Lu et al. (2020)
CO_2-responsive cellulose nanofibril aerogel	Methyl blue Naphthol green B Methyl orange	Yang et al. (2021)
ZnO-$CdWO_4$ nanoparticles	Congo Red	Fatima et al. (2021)
Sugarcane bagasse impregnated with iron oxides	Methylene blue, Malachite green. Reactive Red 535 Remazol Brilliant Blue R	Buttiyappan et al. (2019)

Dyes

(Continued)

TABLE 4.2 (CONTINUED)

Studies on the Removal of a Wide Variety of Different Types of Pollutants Using Various Green Adsorbents

Adsorbent	Pollutants	References
Date seed active carbon (DSAC)	Acid yellow 99	Al-Wasidi et al. (2022)
Cucumber peel-activated carbon/metal-organic framework composites	Acid Green 25 Reactive Yellow 186	Mahmoodi et al. (2019)
Date seed active carbon (DSAC)	Malachite green	El-Bindary et al. (2022)
Alginate-whey	Crystal violet	Djelad et al. (2019)
Starch-based hydrogel	Malachite green	Al-Aidy and Amdeha (2021)
Chitin nanowhiskers (CNW)	Carmine dye	Meshkat et al. (2019)
PVA-co-poly(methacrylic acid)	Methylene blue	Kaur et al. (2020)
$Cu_{0.5}Zn_{0.5}Fe_2O_4$ nanoparticles	Reactive Blue 222	Yeganeh et al. (2020)
Gellan gum/bacterial cellulose (GB) hydrogels	Safranin and crystal violet dye	Nguyen et al. (2022)
SiO_2 nanostructures	Acid Orange 8	Sharma et al. (2022)
Raw sunflower seed shell particles (RSSSP)	Disperse Red 1 Crystal violet Malachite green Methylene blue	Mousavi et al. (2022)
$CoFe_2O_4$@ Methylcellulose (MC)/activated carbon (AC)	Reactive Red 198	Nasiri et al. (2022)
Montmorillonite-reduced graphene oxide composite aerogel (M-rGO)	Methylene blue	Zhou et al. (2022)
CuO nanoparticles CuO/Fe_3O_4 nanocomposite	Methyl orange (MO)	Golabiazar et al. (2022)
Fe^0@chitosan/cellulose composite	Malachite green	Abukhadra et al. (2022)
Mn_3O_4 nanoparticles	Malachite green	Van Tran et al. (2022)
ZnO, CuO, MnO_2, and MgO-based nanoadsorbent	Reactive Golden Yellow-145, Direct Red-3	Munir et al. (2023)
Chitosan-based beads modified with choline chloride:urea Deep eutectic solvent and FeO	Acid Blue 80	Blanco et al. (2023)

cleaning technologies to ensure that all emissions comply with the relevant limits. Therefore, it is possible to identify two main directions in catalysis research in sustainable green chemistry. A revolutionary or long-term approach relies on innovation in catalytic materials chemistry and the use of unconventional reaction conditions that can provide the impetus to develop new solutions, which can create catalysts, technologies, and processes. An evolutionary or short- to medium-term approach mainly focused on improving existing catalytic technologies and processes (Centi & Perathoner 2003).

4.2.5 SURFACTANTS IN GREEN CHEMISTRY

The surfactant, surface active agent, consists of polar and nonpolar parts, and the polar head is hydrophilic, while the tail, usually consisting of a branched, aromatic or linear hydrocarbon chain, is hydrophobic. This amphiphilic structure allows it to reduce surface tension at liquid and gas, liquid and solid, and liquid–liquid interfaces (Pletnev 2001). Among the surfactant applications, products such as household, medicine, food, textile, and cosmetics are predominant. Therefore, it can be said that environmental pollution caused by surfactants increases due to these common practices in daily life (Eng et al. 2006). The availability and variable costs of raw materials, coupled with increasing user interest in lightweight, sustainable, and environmentally friendly ingredients, is increasing the industrial appeal of bio-based surfactants (Foley et al. 2012; Patel 2003; Kokel & Török 2018; Bhadani et al. 2020). The driving forces behind the development of bio-based surfactants, especially those related to raw materials, overlap with many of the principles of green chemistry. However, in order to be sustainable and reduce the negative impact of the industry on the environment, process, and chemical design must also consider factors beyond the raw material source (Stubbs et al. 2022). Surfactants with monosaccharide head groups are the most popular as they can be more stable than those of petrochemical origin and can sufficiently lower the surface tension at low concentrations. Biocompatibility and high biodegradability are also important properties of monosaccharide-derived surfactants. Head groups from glucose and sucrose, as well as alkyl polyglucosides and sucrose esters, are the main surfactants containing monosaccharides produced industrially in the personal care products industry (Manko & Zdziennicka 2015).

4.2.5.1 Biosorbents

As better surfactants can be produced with advanced technology and innovation in parallel with the continuous development in the surfactant industry, studies are either class-specific or product-specific due to the nature of their assessment. In the synthesis of any bio-based product, the main cost-determining factor is the choice of raw materials. Interest in using waste and residues from other industries as chemical raw materials has increased significantly, especially since they can be alternatives to fossil-derived products without the need for additional production systems or land use (Lokesh et al. 2019).

Unlike most surfactants, alkyl polyglucosides, which are mainly synthesized from oleo-chemicals, that is, from plant-derived raw materials, belong to the group of nonionic surfactants. Beginning in the 1960s, the use of petrochemical surfactants began to decline due to their harmful effects on the environment and especially on aquatic ecosystems, and aesthetic concerns such as foaming of surface water, and therefore oleo-chemical surfactants were used primarily in most personal care products (Guilbot et al. 2013).

Polyglucosides are used more in detergents and hard surface cleaners, laundry detergents, softeners, and personal care products such as conditioners and shampoos, due to their emulsifying, foaming, and wetting properties at relatively lower temperatures compared to their counterparts (Lokesh et al. 2017). However, although polyglucosides are bio-based and considered relatively better for the environment compared to synthetic surfactants, they are not emission-free, and especially the material and energy inputs used for their production, as well as the process waste outputs, can cause some adverse environmental impacts. Currently, polyglucosides are synthesized via Fischer glycosidation, which involves an acid-catalyzed reaction between vegetable-based fatty alcohols, typically from palm or coconut oil, and carbohydrates, such as glucose, usually obtained from the hydrolysis of refined syrup or corn starch. All of these raw materials are obtained by human labor and their cultivation requires inputs of water, fertilizers, pesticides, harvesting, transportation, and other energy and chemically intensive activities that have the potential to produce environmental pollution (Brière et al. 2018). Polyglucosides are considered as environmentally friendly surfactants due to their advantages of being low irritant, nontoxic, and easily degradable in the environment, as well as having the advantages of conventional alkyl glycoside surfactants (Zhang et al. 2011).

4.2.5.2 Sucrose esters

Sucrose esters or sucrose fatty acid esters, synthesized from sucrose and fatty acids or glycerides through chemically transesterification reactions, biocatalysis, and reaction with acid halides, are a group of unnatural surfactants. This group of substances is notable for its wide hydrophilic–lipophilic balance (HLB) range. The long fatty acid chain functions as a lipophilic part of the molecule while the polar sucrose moiety functions as a hydrophilic part of the molecule. Because of this amphipathic property, sucrose esters act as emulsifiers; that is, they have the ability to bind both water and oil at the same time. Depending on the HLB value, some can be used as water-in-oil emulsifiers and some as oil-in-water emulsifiers. Due to the wide range of HLB values, they can increase or decrease drug release or absorption (Szűts et al. 2010). Olestra, a class of sucrose esters with highly substituted hydroxyl groups, is also used as a fat replacer in foods (Vargas et al. 2020; Kürti et al. 2012; Szűts & Szabó-Révész 2012; Todosijević et al. 2014).

Sucrose esters, which can be used in cosmetics, food preservatives, and food additives, as well as in products such as emulsifiers, stabilizers, detergents, and oral care agents, are potential biological materials compatible with antibacterial agents based on pharmaceutical applications, in addition to being excellent non-ionic surfactants. (Neta et al. 2015; Moldes et al. 2021; Hayes & Smith 2019; Otache et al. 2022).

Methods for the synthesis of sucrose esters include the use of supercritical carbon dioxide, ionic liquids and ultrasound, pervaporation dehydration, water removal by molecular sieves, immobilized *Rhizomucor miehei* lipase-catalyzed esterification, recombinant DNA technology (Habulin et al. 2008; Xiao et al. 2005; Sakaki et al. 2006). ;Yoo et al. 2007; Ye et al. 2010; Cheng & Shaw 2009).

As the modern pharmaceutical industry shows an increasing interest in biomaterials and green technology, new types of heat-sensitive, nontoxic biomacromolecules are gaining importance. Because of their nontoxic, biodegradable, and gel-forming properties, sucrose esters are also attracting attention as promising substances for creating heat-sensitive delivery systems (Szűts et al. 2010).

4.2.6 INNOVATIVE AGRICULTURAL TECHNOLOGIES BASED ON GREEN CHEMISTRY

Today, the agricultural sector is faced with a variety of global challenges and environmental problems such as climate change, urbanization, unsustainable use of resources, and the access and accumulation of excessive pesticides and fertilizers in the environment (Villaseñor & Ríos 2018). With the use of fertilizers and pesticides, much more people can be fed by using much less land, but these two products also cause serious environmental problems. While agricultural wastes lead to the eutrophication of water bodies, the use of broad-spectrum pesticides can harm beneficial organisms as well as target pests, thus making use of green chemistry much more essential for this field (Kirchhoff 2005; Perlatti et al. 2014).

4.2.6.1 Biorefineries

A biorefinery is a facility where different low-value renewable biomass materials are the raw material for processes where they are converted into higher-value bio-based products such as fibers, food, feed, fine chemicals, transportation fuels and heat, through multiple stages including fractionation, separation, and conversion (Audsley & Annetts 2003). Each refining stage is also called a cascade stage. Using biomass as a raw material can result in lower pollutant emissions and reduced emissions of hazardous products, thereby reducing negative impacts on the environment. A biorefinery can be created by a single unit or combine several facilities targeted for a single purpose, which further processes products as well as by-products or waste from the combined facilities (Mikkola et al. 2015). The biorefinery also has additional objectives such as providing new building blocks for the production of new materials with properties, creating new jobs, including in rural areas, utilizing agricultural, urban and industrial waste, and achieving the ultimate goal of reducing greenhouse gas emissions. The International Energy Agency Bioenergy Task 42 defines biorefining as the sustainable processing of biomass into the bioenergy spectrum in the form of bio-based products such as food, feed, chemicals, materials and biofuels, power, and/or heat. To be able to satisfactorily declare that biorefinery products are green and sustainable, it is necessary to overcome the challenge of using green chemical technologies to ensure maximum recycling efficiency and minimum waste. The emerging biorefinery concept is similar to today's oil refineries, but mostly focuses on specific technologies and raw materials such as starch or vegetable oils that can compete with food or feed (Luque et al. 2008). Therefore, it is necessary to build zero-waste biorefineries that are flexible enough to accept a variety of low-value local raw materials and compete with existing industries. The main biorefinery technologies can be classified as extraction, biochemical, and thermochemical processes. The integrated, near-zero-waste system can use a sequential extraction process that follows a combination of biochemical and thermal processing involving energy and internal recycling of waste gases. Also, removing valuable chemicals before they are destroyed during biochemical and thermochemical can significantly increase overall financial returns (Clark et al. 2012). Depending on the starting material, it is possible to develop many different biorefinery concepts. Currently, there are three main plants under development: whole crop biorefineries based on grain or maize, lignocellulose feedstock biorefineries using dry cellulose-containing biomass or waste, and a green biorefinery using wet biomass such as grass, lucerne, alfalfa, and clover, (Kamm & Kamm 2004; Ryan & Senge 2015).

Biorefineries can be divided into three categories, based on biomass chemistry, such as sugar and starch, lignocellulose, and triglyceride biorefineries.

4.2.6.2 Sugar and starch biorefineries

This type of biorefinery can use various sucrose-containing feedstocks such as sugar beet and sugarcane and starchy biomass such as corn, wheat, barley, and maize to produce bioethanol. Bioethanol is an important biofuel mixed with gasoline, but is also used for the synthesis of chemicals such as diethyl ether and ethylene.

4.2.6.3 Lignocellulosic biorefinery

This biorefinery uses natural dry biomass such as cellulosic biomass and agricultural wastes to produce biofuels and other bioproducts (Maity 2015).

4.2.6.4 Triglyceride (oil and fats) biorefinery

This biorefinery uses vegetable and animal oils to produce biodiesel with properties comparable to petrodiesel. Also, the production of biodiesel as fatty acid methyl ester is based on the transesterification of triglycerides with methanol. The reaction is usually catalyzed by acid, alkali, or enzymes, depending on the free fatty acids content of the feedstock (Mikkola et al. 2015).

4.2.6.5 Bioethanol

Bioethanol, which is promising in tackling today's global energy crisis and deteriorating environmental quality and is currently the most widely used liquid biofuel for motor vehicles, is a renewable and sustainable liquid fuel (Aditiya et al. 2016). The importance of ethanol has gradually increased due to a number of reasons such as global warming and climate change, and thus bioethanol has entered the stage of attracting wide attention at the global level and rapidly growing global market (Sarkar et al. 2012). The desire and effort to achieve less dependence on fossil resources has been the main driving force behind the significant increase in the amount of bioethanol produced by the fermentation of biomass (Rass-Hansen et al. 2007). Although it has a two-thirds lower energy content compared to petroleum, the higher oxygen content of bioethanol makes combustion cleaner and leads to lower toxic substance emissions (Krylova et al. 2008). Thus, bioethanol, which can reduce CO_2 emissions by up to 80% compared to gasoline, supports a cleaner environment for the future. The criteria for bioethanol to replace or mix with gasoline are specified in the ASTM D4806 standard, which determines the quality requirements of bioethanol for spark ignition engines (Sebayang et al. 2016). The quality of the bioethanol produced is highly dependent on the production routes, and as bioethanol production, in general, consists of several sequential procedures such as pretreatment, hydrolysis, fermentation, and distillation, each of the stages is branched and each branch has different effects on ethanol quality and overall production cost (Aditiya et al. 2016).

4.2.6.6 Biofuels

In parallel with the increasing need for energy globally, the use of fossil fuels, which create undesirable environmental effects, is increasing day by day (Karthikeyan et al. 2017). High population and industrial growth have increased the demand and price of fossil fuels, which has led researchers to find renewable and environmentally friendly fuels (Kumar et al. 2019). Fuel is an inevitable need in transportation and industry, and the global demand for crude oil is currently at the level of 101.6 million barrels/day (Naveen Kumar & Baskar 2020). Biofuels vary according to many basic features such as the type of raw material, the conversion process, and the technical characteristics of the fuel and its use. Because of these many possible differences, there are various definitions for biofuel types. Two commonly used typologies are "first, second, and third generation" and "traditional and advanced" biofuels. Biofuels produced from food or animal feed plants are called first-generation biofuels, but they are also called conventional biofuels because they are produced by known technology and processes such as fermentation, distillation, and transesterification. The main feature of second-generation biofuels is that they are derived from specialized energy crops such as miscanthus or silvergrass, switchgrass, short-rotation coppice and other lignocellulosic plants, and non-food raw materials such as agricultural wastes and forest residues. Generally, biodiesel produced from microalgae by transesterification or hydrotreatment of algae oil is recognized as a third-generation biofuel. Second- and third-generation biofuels are often referred to as advanced biofuels, as their production techniques are still in research and development and partially pilot or demonstration stages (Jeswani et al. 2020). Environmentally friendly and nontoxic renewable fuels such as biodiesel and bioethanol play a very important role in the energy sector. Biodiesel is considered a biodegradable and non-lethal sustainable power source that can be obtained from various sources such as plant, animal fat, and algae oil (Chakraborty et al. 2016). Biodiesel, sometimes called fatty acid methyl ester, contains long chains of mono-alkyl esters containing unsaturated fat formed by transesterification (Naveen Kumar & Baskar 2021). On the other hand, alkaline earth metal oxides such as calcium oxide and calcium methoxide have excellent catalytic activity and stability for biodiesel production (Feyzi & Shahbazi 2015). Strong base soluble hydroxides are substances that can be considered in the transesterification of triglycerides for efficient biodiesel production. Zinc-doped calcium oxide can also provide an increased fatty acid methyl ester or biodiesel production yield (Istadi et al. 2015; Outili et al. 2020).

4.2.7   ENERGY CONVERSION AND STORAGE SYSTEMS BASED ON GREEN CHEMISTRY

The social welfare and economic development of the modern world are directly related to sustainable energy conversion and storage, and one of the biggest challenges for a sustainable society is energy supply (Arico et al. 2005; Chu & Majumdar 2012; Zhang et al. 2021). Energy scarcity and environmental pollution caused by insufficient fossil fuel resources and increased consumption have become two major global problems of humans (Chu & Majumdar 2012). Developing new technology to take full advantage of abundant green energies in the form of solar, mechanical, and thermal energies appears to be a promising and effective way to meet long-term energy needs and environmentally sustainable development. In the last few decades, numerous energy conversion technologies such as solar cells, mechanical generators, and thermoelectric generators have been developed to convert green energies into electrical energy, which is the most widely used type of energy today (Luo & Wang 2017).

4.2.7.1   Solar cells

Solar energy has much greater potential as a clean and sustainable energy source, as it is a continuous and inexhaustible resource beyond other renewable energy sources in nature (Wijewardane 2009). Solar cells that convert light energy directly into electricity with the photovoltaic effect are defined as photovoltaic regardless of whether the source is the sun or artificial light. Besides generating energy, they can be used as photodetectors such as infrared detectors to detect light or other electromagnetic radiation near the visible distance or to measure light intensity. For a solar cell or photovoltaic cell to work, it requires three basic properties: absorbing light, producing excitons (bonded electron-hole pairs), unbonded electron-hole pairs (via excitons), or plasmons. New-generation solar cells based on thin-film technology include dye-sensitized solar cells, quantum dot-sensitized solar cells, and polymer/organic solar cells, which are promising photovoltaic devices due to their low production costs and high energy conversion performances (Tan et al. 2012). A photoelectrolytic cell, on the other hand, can refer either to a photovoltaic cell, such as a type of dye-sensitized solar cell, or to a system that uses only sunlight to separate water directly into oxygen and hydrogen.

4.2.7.2 Hydrogen storage

Hydrogen storage can be accomplished by a variety of methods, including mechanical approaches such as retaining hydrogen for later use, using existing high pressures and low temperatures, or using chemical compounds that release hydrogen as needed. It produces large quantities of hydrogen from various industries, and the hydrogen produced is mostly consumed for the production of ammonia. For many years, hydrogen stored as compressed gas or cryogenic liquid has been transported as such in cylinders, tubes, and cryogenic tanks for use as propellants in industry or space programs. Interest in using hydrogen for onboard energy storage in zero-emission vehicles motivates the development of new storage methods. However, hydrogen's very low boiling point of −252.8°C and thus the difficulty of reaching such low temperatures is a serious problem and requires significant energy expenditure. Storage of hydrogen as a gas, which can be physically stored as a gas or liquid, typically requires high-pressure tanks (350–700 bar). Nanocomposites appear to be promising sorbents for hydrogen storage, which is seen as a new clean energy source. The strategies identified are generally based on hydrogen entrapment in microporous media with surface area, which stands out as carbon-based, metal-organic frameworks, and dual-phase nanocomposite materials (Sculley et al. 2011; Choucair & Mauron 2015; Yadav et al. 2015).

REFERENCES

Abdussalam-Mohammed, W., Ali, A. Q., & Errayes, A. O. (2020). Green chemistry: Principles, applications, and disadvantages. *Chemical Methodologies*, 4(4), 408–423.

Abou-Shehada, S., Clark, J. H., Paggiola, G., & Sherwood, J. (2016). Tunable solvents: Shades of green. *Chemical Engineering and Processing: Process Intensification*, 99, 88–96.

Abukhadra, M. R., Saad, I., Othman, S. I., Katowah, D. F., Ajarem, J. S., Alqarni, S. A., Allam, A. A., Zoubi, W. A., & Ko, Y. G. (2022). Characterization of Fe⁰@chitosan/cellulose structure as effective green adsorbent for methyl parathion, malachite Green, and levofloxacin removal: Experimental and theoretical studies. *Journal of Molecular Liquids*, 368, 120730.

Adeleye, A. T., Akande, A. A., Odoh, C. K., Philip, M., Fidelis, T. T., Amos, P. I., & Banjoko, O. O. (2021). Efficient synthesis of bio-based activated carbon (AC) for catalytic systems: A green and sustainable approach. *Journal of Industrial and Engineering Chemistry*, 96, 59–75.

Aditiya, H. B., Mahlia, T. M. I., Chong, W. T., Nur, H., & Sebayang, A. H. (2016). Second generation bioethanol production: A critical review. *Renewable and Sustainable Energy Reviews*, 66, 631–653.

Ahsan, M. A., Jabbari, V., Islam, M. T., Kim, H., Hernandez-Viezcas, J. A., Lin, Y., Díaz-Moreno, C. A., Lopez, J., Gardea-Torresdey, J., & Noveron, J. C. (2018). Green synthesis of a highly efficient biosorbent for organic, pharmaceutical, and heavy metal pollutants removal: Engineering surface chemistry of polymeric biomass of spent coffee waste. *Journal of Water Process Engineering*, 25, 309–319.

Ahuja, S. (2013). Green chemistry and other novel solutions to water pollution: Overview. In *Novel solutions to water pollution*, Satinder, A., & Kiril, H., (Eds), (pp. 1–14). Oxford University Press.

Ajmal, M., Khan Rao, R. A., & Ahmad, R. (2011). Adsorption studies of heavy metals on Tectona grandis: Removal and recovery of Zn (II) from electroplating wastes. *Journal of Dispersion Science and Technology*, 32(6), 851–856.

Akiya, N., & Savage, P. E. (2002). Roles of water for chemical reactions in high-temperature water. *Chemical Reviews*, 102(8), 2725–2750.

Al-Aidy, H., & Amdeha, E. (2021). Green adsorbents based on polyacrylic acid-acrylamide grafted starch hydrogels: The new approach for enhanced adsorption of malachite green dye from aqueous solution. *International Journal of Environmental Analytical Chemistry*, 101(15), 2796–2816.

Ali, H., & Khan, E. (2018). What are heavy metals? Long-standing controversy over the scientific use of the term 'heavy metals'–proposal of a comprehensive definition. *Toxicological and Environmental Chemistry*, 100(1), 6–19.

AlQadhi, N. F., & AlSuhaimi, A. O. (2020). Chemically functionalized activated carbon with 8-hydroxyquinoline using aryldiazonium salts/diazotization route: Green chemistry synthesis for oxins-carbon chelators. *Arabian Journal of Chemistry*, 13(1), 1386–1396.

Al-Wasidi, A. S., AlZahrani, I. I., Thawibaraka, H. I., Naglah, A. M., El-Desouky, M. G., & El-Bindary, M. A. (2022). Adsorption studies of carbon dioxide and anionic dye on green adsorbent. *Journal of Molecular Structure*, 1250, 131736.

Ambika, S., Kumar, M., Pisharody, L., Malhotra, M., Kumar, G., Sreedharan, V., Singh, L., Nidheesh, P. V., & Bhatnagar, A. (2022). Modified biochar as a green adsorbent for removal of hexavalent chromium from various environmental matrices: Mechanisms, methods, and prospects. *Chemical Engineering Journal*, 439, 135716.

Anastas, N. D., & Warner, J. C. (2005). The incorporation of hazard reduction as a chemical design criterion in green chemistry. *Chemical Health and Safety*, 12(2), 9–13.

Anastas, P., & Eghbali, N. (2010). Green chemistry: Principles and practice. *Chemical Society Reviews*, 39(1), 301–312.

Anastas, P. T., Bartlett, L. B., Kirchhoff, M. M., & Williamson, T. C. (2000). The role of catalysis in the design, development, and implementation of green chemistry. *Catalysis Today*, 55(1–2), 11–22.

Anastas, P. T., & Warner, J. (1998). *Green chemistry theory and practice*. Oxford University Press.

Anastopoulos, I., Robalds, A., Tran, H. N., Mitrogiannis, D., Giannakoudakis, D. A., Hosseini-Bandegharaei, A., & Dotto, G. L. (2019). Removal of heavy metals by leaves-derived biosorbents. *Environmental Chemistry Letters*, 17(2), 755–766.

Anfar, Z., Zbair, M., Ahsiane, H. A., Jada, A., & El Alem, N. (2020). Microwave assisted green synthesis of Fe_2O_3/biochar for ultrasonic removal of nonsteroidal anti-inflammatory pharmaceuticals. *RSC Advances*, 10(19), 11371–11380.

Arabkhani, P., & Asfaram, A. (2022). The potential application of bio-based ceramic/organic xerogel derived from the plant sources: A new green adsorbent for removal of antibiotics from pharmaceutical wastewater. *Journal of Hazardous Materials*, 429, 128289.

Ardila-Fierro, K. J., & Hernández, J. G. (2021). Sustainability assessment of mechano-chemistry by using the twelve principles of green chemistry. *ChemSusChem*, *14*(10), 2145–2162.

Arico, A. S., Bruce, P., Scrosati, B., Tarascon, J. M., & Van Schalkwijk, W. (2005). Nanostructured materials for advanced energy conversion and storage devices. *Nature Materials*, *4*(5), 366–377.

Arthur, C. L., & Pawliszyn, J. (1990). Solid phase microextraction with thermal desorption using fused silica optical fibers. *Analytical Chemistry*, *62*(19), 2145–2148.

Audsley, E., & Annetts, J. E. (2003). Modelling the value of a rural biorefinery—Part I: The model description. *Agricultural Systems*, *76*(1), 39–59.

Auffan, M., Rose, J., Bottero, J. Y., Lowry, G. V., Jolivet, J. P., & Wiesner, M. R. (2009). Towards a definition of inorganic nanoparticles from an environmental, health and safety perspective. *Nature Nanotechnology*, *4*(10), 634–641.

Badami, B. V. (2008). Concept of green chemistry. *Resonance*, *13*(11), 1041–1048.

Bagheri, A. R., Aramesh, N., & Haddad, P. R. (2021). Applications of covalent organic frameworks and their composites in the extraction of pesticides from different samples. *Journal of Chromatography: Part A*, *1661*, 462612.

Barkade, S., Sable, S., Ashtekar, V., & Pandit, V. (2022). Removal of lead and copper from wastewater using Bael fruit shell as an adsorbent. *Materials Today: Proceedings*, *53*, 65–70.

Beckman, E. J. (2004). Supercritical and near-critical CO2 in green chemical synthesis and processing. *Journal of Supercritical Fluids*, *28*(2–3), 121–191.

Bedia, J., Peñas-Garzón, M., Gómez-Avilés, A., Rodriguez, J. J., & Belver, C. (2018). A review on the synthesis and characterization of biomass-derived carbons for adsorption of emerging contaminants from water. *C*, *4*(4), 63.

Bhadani, A., Kafle, A., Ogura, T., Akamatsu, M., Sakai, K., Sakai, H., & Abe, M. (2020). Current perspective of sustainable surfactants based on renewable building blocks. *Current Opinion in Colloid and Interface Science*, *45*, 124–135.

Bhandari, M. E. E. N. A., & Raj, S. E. E. M. A. (2017). Practical approach to green chemistry. *International Journal of Pharmacy and Pharmaceutical Sciences*, *9*(4), 10–26.

Blackmond, D. G., Armstrong, A., Coombe, V., & Wells, A. (2007). Water in organocatalytic processes: Debunking the myths. *Angewandte Chemie International Edition*, *46*(21), 3798–3800.

Blanco, L., Martínez-Rico, O., Domínguez, Á., & González, B. (2023). Removal of Acid Blue 80 from aqueous solutions using chitosan-based beads modified with choline chloride: Urea deep eutectic solvent and FeO. *Water Resources and Industry*, *29*, 100195.

Boruah, P. K., Sharma, B., Hussain, N., & Das, M. R. (2017). Magnetically recoverable Fe_3O_4/graphene nanocomposite towards efficient removal of triazine pesticides from aqueous solution: Investigation of the adsorption phenomenon and specific ion effect. *Chemosphere*, *168*, 1058–1067.

Boyère, C., Jérôme, C., & Debuigne, A. (2014). Input of supercritical carbon dioxide to polymer synthesis: An overview. *European Polymer Journal*, *61*, 45–63.

Brière, R., Loubet, P., Glogic, E., Estrine, B., Marinkovic, S., Jérôme, F., & Sonnemann, G. (2018). Life cycle assessment of the production of surface-active alkyl polyglycosides from acid-assisted ball-milled wheat straw compared to the conventional production based on corn-starch. *Green Chemistry*, *20*(9), 2135–2141.

Briffa, J., Sinagra, E., & Blundell, R. (2020). Heavy metal pollution in the environment and their toxicological effects on humans. *Heliyon*, *6*(9), e04691.

Brunner, G. (2005). Supercritical fluids: Technology and application to food processing. *Journal of Food Engineering*, *67*(1–2), 21–33.

Buthiyappan, A., Gopalan, J., & Raman, A. A. A. (2019). Synthesis of iron oxides impregnated green adsorbent from sugarcane bagasse: Characterization and evaluation of adsorption efficiency. *Journal of Environmental Management*, *249*, 109323.

Byrne, F. P., Jin, S., Paggiola, G., Petchey, T. H., Clark, J. H., Farmer, T. J., Hunt, A. J., McElroy, C. R., & Sherwood, J. (2016). Tools and techniques for solvent selection: Green solvent selection guides. *Sustainable Chemical Processes*, *4*(1), 1–24.

Centi, G., & Perathoner, S. (2003). Catalysis and sustainable (green) chemistry. *Catalysis Today*, *77*(4), 287–297.

Chakraborty, R., Chatterjee, S., Mukhopadhyay, P., & Barman, S. (2016). Progresses in waste biomass derived catalyst for production of biodiesel and bioethanol: A review. *Procedia Environmental Sciences*, *35*, 546–554.

Chandrasekhar, S., Narsihmulu, C., Sultana, S. S., & Reddy, N. R. (2002). Poly (ethylene glycol)(PEG) as a reusable solvent medium for organic synthesis. Application in the Heck reaction. *Organic Letters*, *4*(25), 4399–4401.

Chang, S. W., & Shaw, J. F. (2009). Biocatalysis for the production of carbohydrate esters. *New Biotechnology*, *26*(3–4), 109–116.

Chatzimitakos, T., Binellas, C., Maidatsi, K., & Stalikas, C. (2016). Magnetic ionic liquid in stirring-assisted drop-breakup microextraction: Proof-of-concept extraction of phenolic endocrine disrupters and acidic pharmaceuticals. *Analytica Chimica Acta*, *910*, 53–59.

Chen, T. L., Kim, H., Pan, S. Y., Tseng, P. C., Lin, Y. P., & Chiang, P. C. (2020). Implementation of green chemistry principles in circular economy system towards sustainable development goals: Challenges and perspectives. *Science of the Total Environment*, *716*, 136998.

Chen, Y., Liu, H., Xia, M., Cai, M., Nie, Z., & Gao, J. (2023). Green multifunctional PVA composite hydrogel-membrane for the efficient purification of emulsified oil wastewater containing Pb2+ ions. *Science of the Total Environment*, *856*(2), 159271.

Chijioke, J. A., Esther, O. F., Chijioke, J. A., & Esther, O. F. (2021). Adsorption of copper (II) ions onto raw Globimetula oreophila (Afomo ori koko) leaves. *African Journal of Biotechnology*, *20*(3), 122–133.

Chojnacka, K. (2010). Biosorption and bioaccumulation–the prospects for practical applications. *Environment International*, *36*(3), 299–307.

Choucair, M., & Mauron, P. (2015). Versatile preparation of graphene-based nanocomposites and their hydrogen adsorption. *International Journal of Hydrogen Energy*, *40*(18), 6158–6164.

Chowdhury, A., Khan, A. A., Kumari, S., & Hussain, S. (2019). Superadsorbent Ni–Co–S/SDS nanocomposites for ultrahigh removal of cationic, anionic organic dyes and toxic metal ions: Kinetics, isotherm and adsorption mechanism. *ACS Sustainable Chemistry and Engineering*, *7*(4), 4165–4176.

Chu, S., & Majumdar, A. (2012). Opportunities and challenges for a sustainable energy future. *Nature*, *488*(7411), 294–303.

Clark, J. H., Farmer, T. J., Hunt, A. J., & Sherwood, J. (2015). Opportunities for bio-based solvents created as petrochemical and fuel products transition towards renewable resources. *International Journal of Molecular Sciences*, *16*(8), 17101–17159.

Clark, J. H., Luque, R., & Matharu, A. S. (2012). Green chemistry, biofuels, and biorefinery. *Annual Review of Chemical and Biomolecular Engineering*, *3*(1), 183–207.

Clark, J. H., & Tavener, S. J. (2007). Alternative solvents: Shades of green. *Organic Process Research and Development*, *11*(1), 149–155.

Clarke, C. J., Tu, W. C., Levers, O., Brohl, A., & Hallett, J. P. (2018). Green and sustainable solvents in chemical processes. *Chemical Reviews*, *118*(2), 747–800.

Colozza, N., Caratelli, V., Moscone, D., & Arduini, F. (2021). Based devices as new smart analytical tools for sustainable detection of environmental pollutants. *Case Studies in Chemical and Environmental Engineering, 4*, 100167.

Constable, D. J., Jimenez-Gonzalez, C., & Henderson, R. K. (2007). Perspective on solvent use in the pharmaceutical industry. *Organic Process Research and Development, 11*(1), 133–137.

Cuartero, M., Colozza, N., Fernández-Pérez, B. M., & Crespo, G. A. (2020). Why ammonium detection is particularly challenging but insightful with ionophore-based potentiometric sensors–an overview of the progress in the last 20 years. *Analyst, 145*(9), 3188–3210.

Cvjetko Bubalo, M., Vidović, S., Radojčić Redovniković, I., & Jokić, S. (2015). Green solvents for green technologies. *Journal of Chemical Technology and Biotechnology, 90*(9), 1631–1639.

Dąbrowski, A. (2001). Adsorption—From theory to practice. *Advances in Colloid and Interface Science, 93*(1–3), 135–224.

Dallinger, D., & Kappe, C. O. (2007). Microwave-assisted synthesis in water as solvent. *Chemical Reviews, 107*(6), 2563–2591.

Dan, Y., Xu, L., Qiang, Z., Dong, H., & Shi, H. (2021). Preparation of green biosorbent using rice hull to preconcentrate, remove and recover heavy metal and other metal elements from water. *Chemosphere, 262*, 127940.

Darweesh, M. A., Elgendy, M. Y., Ayad, M. I., Ahmed, A. M. M., Elsayed, N. K., & Hammad, W. A. (2022). A unique, inexpensive, and abundantly available adsorbent: Composite of synthesized silver nanoparticles (AgNPs) and banana leaves powder (BLP). *Heliyon, 8*(4), e09279.

de Godoi Pereira, M., & Arruda, M. A. Z. (2003). Trends in preconcentration procedures for metal determination using atomic spectrometry techniques. *Microchimica Acta, 141*(3), 115–131.

del Pilar Sánchez-Camargo, A., Bueno, M., Parada-Alfonso, F., Cifuentes, A., & Ibáñez, E. (2019). Hansen solubility parameters for selection of green extraction solvents. *TrAC Trends in Analytical Chemistry, 118*, 227–237.

Dercová, K., Makovníková, J., Barančíková, G., & Žuffa, J. (2005). Bioremediation of soil and wastewater contaminated with toxic metals. *Chemické Listy, 99*(10), 682–693.

Dhankhar, R., & Hooda, A. (2011). Fungal biosorption–an alternative to meet the challenges of heavy metal pollution in aqueous solutions. *Environmental Technology, 32*(5), 467–491.

Diel, J. C., Franco, D. S., Nunes, I. D. S., Pereira, H. A., Moreira, K. S., Burgo, T. A. D. L., Foletto, E. L. & Dotto, G. L. (2021). Carbon nanotubes impregnated with metallic nanoparticles and their application as an adsorbent for the glyphosate removal in an aqueous matrix. *Journal of Environmental Chemical Engineering, 9*(2), 105178.

Divyashree, A., & Hegde, G. (2015). Activated carbon nanospheres derived from bio-waste materials for supercapacitor applications–a review. *RSC Advances, 5*(107), 88339–88352.

Djelad, A., Mokhtar, A., Khelifa, A., Bengueddach, A., & Sassi, M. (2019). Alginate-whey an effective and green adsorbent for crystal violet removal: Kinetic, thermodynamic and mechanism studies. *International Journal of Biological Macromolecules, 139*, 944–954.

Dua, R., Shrivastava, S., Shrivastava, S. L., & Srivastava, S. K. (2012). Green Chemistry and environmentally friendly technologies: A review. *Middle-East Journal of Scientific Research, 11*(7), 846–855.

Dube, M. A., & Salehpour, S. (2014). Applying the principles of green chemistry to polymer production technology. *Macromolecular Reaction Engineering, 8*(1), 7–28.

Earle, M. J., & Seddon, K. R. (2000). Ionic liquids: Green solvents for the future. *Pure and Applied Chemistry*, *72*(7), 1391–1398.

El-Araby, H. A., Ibrahim, A. M. M. A., Mangood, A. H., & Adel, A. H. (2017). Sesame husk as adsorbent for copper (II) ions removal from aqueous solution. *Journal of Geoscience and Environment Protection*, *5*(7), 109.

El-Bindary, M. A., El-Desouky, M. G., & El-Bindary, A. A. (2022). Adsorption of industrial dye from aqueous solutions onto thermally treated green adsorbent: A complete batch system evaluation. *Journal of Molecular Liquids*, *346*, 117082.

Elhussien, M. H., Hussein, R. M., Nimir, S. A., & Elsaim, M. H. (2017). Preparation and characterization of activated carbon from palm tree leaves impregnated with zinc chloride for the removal of lead (II) from aqueous solutions. *American Journal of Physical Chemistry*, *6*(4), 59–69.

Elmorsi, R. R., El-Wakeel, S. T., Shehab El-Dein, W. A., Lotfy, H. R., Rashwan, W. E., Nagah, M., Shaaban, S. A., Sayed Ahmed, S. A., El-Sherif, I. Y., & Abou-El-Sherbini, K. S. (2019). Adsorption of methylene blue and Pb2+ by using acid-activated Posidonia oceanica waste. *Scientific Reports*, *9*(1), 1–12.

Eng, Y. Y., Sharma, V. K., & Ray, A. K. (2006). Ferrate (VI): Green chemistry oxidant for degradation of cationic surfactant. *Chemosphere*, *63*(10), 1785–1790.

Erythropel, H. C., Zimmerman, J. B., de Winter, T. M., Petitjean, L., Melnikov, F., Lam, C. H., Lounsburyi, A. W., Mellor, K. E., Janković, N. Z., Tu, Q., Pincus, L. N., Falinski, M. M., Shi, W., Coish, P., Plata, D. L., & Anastas, P. T. (2018). The Green ChemisTREE: 20 years after taking root with the 12 principles. *Green Chemistry*, *20*(9), 1929–1961.

Essebaai, H., Lgaz, H., Alrashdi, A. A., Habsaoui, A., Lebkiri, A., Marzak, S., & Rifi, E. H. (2022). Green and eco-friendly montmorillonite clay for the removal of Cr (III) metal ion from aqueous environment. *International Journal of Environmental Science and Technology*, *19*(4), 2443–2454.

Faber, K. (2011). *Biotransformations in organic chemistry: A textbook* (No. 660.634 F334B.). Springer.

Fatima, B., Siddiqui, S. I., Nirala, R. K., Vikrant, K., Kim, K. H., Ahmad, R., & Chaudhry, S. A. (2021). Facile green synthesis of ZnO–CdWO$_4$ nanoparticles and their potential as adsorbents to remove organic dye. *Environmental Pollution*, *271*, 116401.

Feng, Z., Huang, C., Guo, Y., Liu, W., & Zhang, L. (2020). Graphitic carbon nitride derivative with large mesopores as sorbent for solid-phase microextraction of polycyclic aromatic hydrocarbons. *Talanta*, *209*, 120541.

Feu, K. S., Alexander, F., Silva, S., de Moraes Junior, M. A., Corrêa, A. G., & Paixão, M. W. (2014). Polyethylene glycol (PEG) as a reusable solvent medium for an asymmetric organocatalytic Michael addition. Application to the synthesis of bioactive compounds. *Green Chemistry*, *16*(6), 3169–3174.

Feyzi, M., & Shahbazi, E. (2015). Catalytic performance and characterization of Cs–Ca/SiO$_2$–TiO$_2$ nanocatalysts for biodiesel production. *Journal of Molecular Catalysis A: Chemical*, *404*, 131–138.

Foley, P., Beach, E. S., & Zimmerman, J. B. (2012). Derivation and synthesis of renewable surfactants. *Chemical Society Reviews*, *41*(4), 1499–1518.

Foster, R., Ghassemi, M., & Cota, A. (2009). *Solar energy: Renewable energy and the environment*. CRC Press.

Gao, G., Chen, H., Dai, J., Jin, L., Chai, Y., Zhu, L., Liu, X., & Lu, C. (2020). Determination of polychlorinated biphenyls in tea using gas chromatography–tandem mass spectrometry combined with dispersive solid phase extraction. *Food Chemistry*, *316*, 126290.

Ge, Y., & Li, Z. (2018). Application of lignin and its derivatives in adsorption of heavy metal ions in water: A review. *ACS Sustainable Chemistry and Engineering*, 6(5), 7181–7192.

Ghiasi, A., Malekpour, A., & Mahpishanian, S. (2020). Metal-organic framework MIL101 (Cr)-NH_2 functionalized magnetic graphene oxide for ultrasonic-assisted magnetic solid phase extraction of neonicotinoid insecticides from fruit and water samples. *Talanta*, 217, 121120.

Gindaba, G. T., Demsash, H. D., & Jayakumar, M. (2023). Green synthesis, characterization, and application of metal oxide nanoparticles for mercury removal from aqueous solution. *Environmental Monitoring and Assessment*, 195(1), 1–23.

Godwin, P. M., Pan, Y., Xiao, H., & Afzal, M. T. (2019). Progress in preparation and application of modified biochar for improving heavy metal ion removal from wastewater. *Journal of Bioresources and Bioproducts*, 4(1), 31–42.

Golabiazar, R., Sabr, M. R., Ali, A. A., Qadr, N. S., Rahman, R. S., Othman, K. I., Khalid, K. M., Musa, S. J., & Hamadammin, B. J. (2022). Investigation and characterization of biosynthesized green adsorbent CuO NPs and CuO/Fe_3O_4 NCs using Adiantum CV leaf for removal MO dye and Cr (VI) metal ions: Thermodynamic, kinetic, and antibacterial studies. *Journal of the Iranian Chemical Society*, 19, 3135–3153.

Gouveia, A. S., Oliveira, F. S., Kurnia, K. A., & Marrucho, I. M. (2016). Deep eutectic solvents as azeotrope breakers: Liquid–liquid extraction and COSMO-RS prediction. *ACS Sustainable Chemistry and Engineering*, 4(10), 5640–5650.

Gu, S., Kang, X., Wang, L., Lichtfouse, E., & Wang, C. (2019). Clay mineral adsorbents for heavy metal removal from wastewater: A review. *Environmental Chemistry Letters*, 17(2), 629–654.

Guilbot, J., Kerverdo, S., Milius, A., Escola, R., & Pomrehn, F. (2013). Life cycle assessment of surfactants: The case of an alkyl polyglucoside used as a self emulsifier in cosmetics. *Green Chemistry*, 15(12), 3337–3354.

Habulin, M., Šabeder, S., & Knez, Ž. (2008). Enzymatic synthesis of sugar fatty acid esters in organic solvent and in supercritical carbon dioxide and their antimicrobial activity. *Journal of Supercritical Fluids*, 45(3), 338–345.

Hallett, J. P., & Welton, T. (2011). Room-temperature ionic liquids: Solvents for synthesis and catalysis. *Chemical Reviews*, 111(5), 3508–3576.

Han, X., & Poliakoff, M. (2012). Continuous reactions in supercritical carbon dioxide: Problems, solutions and possible ways forward. *Chemical Society Reviews*, 41(4), 1428–1436.

Hashemi, B., Shamsipur, M., Javadi, A., Rofouei, M. K., Shockravi, A., Tajarrod, N., & Mandumy, N. (2015). Synthesis and characterization of ion imprinted polymeric nanoparticles for selective extraction and determination of mercury ions. *Analytical Methods*, 7(22), 9641–9648.

Hashemi, B., Zohrabi, P., Kim, K. H., Shamsipur, M., Deep, A., & Hong, J. (2017). Recent advances in liquid-phase microextraction techniques for the analysis of environmental pollutants. *TrAC Trends in Analytical Chemistry*, 97, 83–95.

Hayes, D. G., & Smith, G. A. (2019). Biobased surfactants: Overview and industrial state of the art. In *Biobased surfactants* (pp. 3–38). Elsiever.

Heidarinejad, Z., Dehghani, M. H., Heidari, M., Javedan, G., Ali, I., & Sillanpää, M. (2020). Methods for preparation and activation of activated carbon: A review. *Environmental Chemistry Letters*, 18(2), 393–415.

Herrero, M., Mendiola, J. A., Cifuentes, A., & Ibáñez, E. (2010). Supercritical fluid extraction: Recent advances and applications. *Journal of Chromatography: Part A*, 1217(16), 2495–2511.

Horváth, I. T. (2008). Solvents from nature. *Green Chemistry*, 10(10), 1024–1028.

Huang, C., Qiao, X., Sun, W., Chen, H., Chen, X., Zhang, L., & Wang, T. (2019a). Effective extraction of domoic acid from seafood based on postsynthetic-modified magnetic zeolite imidazolate framework-8 particles. *Analytical Chemistry*, *91*(3), 2418–2424.

Huang, X., Qiao, K., Li, L., Liu, G., Xu, X., Lu, R., Gao, H., & Xu, D. (2019b). Preparation of a magnetic graphene/polydopamine nanocomposite for magnetic dispersive solid-phase extraction of benzoylurea insecticides in environmental water samples. *Scientific Reports*, *9*(1), 1–9.

Huang, Z., Chua, P. E., & Lee, H. K. (2015). Carbonized polydopamine as coating for solid-phase microextraction of organochlorine pesticides. *Journal of Chromatography: Part A*, *1399*, 8–17.

Husein, D. Z., Hassanien, R., & Al-Hakkani, M. F. (2019). Green-synthesized copper nano-adsorbent for the removal of pharmaceutical pollutants from real wastewater samples. *Heliyon*, *5*(8), e02339.

Hussain, D., Khan, S. A., Alharthi, S. S., & Khan, T. A. (2022). Insight into the performance of novel kaolinite-cellulose/cobalt oxide nanocomposite as green adsorbent for liquid phase abatement of heavy metal ions: Modelling and mechanism. *Arabian Journal of Chemistry*, *15*(7), 103925.

Ibisi, N. E., & Asoluka, C. A. (2018). Use of agro-waste (*Musa paradisiaca* peels) as a sustainable biosorbent for toxic metal ions removal from contaminated water. *Chemistry International*, *4*(1), 52.

Ighalo, J. O., Omoarukhe, F. O., Ojukwu, V. E., Iwuozor, K. O., & Igwegbe, C. A. (2022). Cost of adsorbent preparation and usage in wastewater treatment: A review. *Cleaner Chemical Engineering*, *3*, 100042.

Iqbal, M., Saeed, A., & Kalim, I. (2009). Characterization of adsorptive capacity and investigation of mechanism of Cu2+, Ni2+ and Zn2+ adsorption on mango peel waste from constituted metal solution and genuine electroplating effluent. *Separation Science and Technology*, *44*(15), 3770–3791.

Islam, M. S., Ahmed, M. K., Raknuzzaman, M., Habibullah-Al-Mamun, M., & Islam, M. K. (2015). Heavy metal pollution in surface water and sediment: A preliminary assessment of an urban river in a developing country. *Ecological Indicators*, *48*, 282–291.

Istadi, I., Prasetyo, S. A., & Nugroho, T. S. (2015). Characterization of K_2O/CaO-ZnO catalyst for transesterification of soybean oil to biodiesel. *Procedia Environmental Sciences*, *23*, 394–399.

Ivanković, A., Dronjić, A., Bevanda, A. M., & Talić, S. (2017). Review of 12 principles of green chemistry in practice. *International Journal of Sustainable and Green Energy*, *6*(3), 39–48.

Jamshidi, F., Nouri, N., Sereshti, H., & Aliabadi, M. H. S. (2020). Synthesis of magnetic poly (acrylic acid-menthol deep eutectic solvent) hydrogel: Application for extraction of pesticides. *Journal of Molecular Liquids*, *318*, 114073.

Jayan, N., Bhatlu M, L. D., & Akbar, S. T. (2021). Central composite design for adsorption of Pb (II) and Zn (II) metals on PKM-2 Moringa oleifera leaves. *ACS Omega*, *6*(39), 25277–25298.

Jessop, P. G. (2011). Searching for green solvents. *Green Chemistry*, *13*(6), 1391–1398.

Jessop, P. G., & Subramaniam, B. (2007). Gas-expanded liquids. *Chemical Reviews*, *107*(6), 2666–2694.

Jeswani, H. K., Chilvers, A., & Azapagic, A. (2020). Environmental sustainability of biofuels: A review. *Proceedings of the Royal Society Series A*, *476*(2243), 20200351.

Ji, L., Chen, W., Duan, L., & Zhu, D. (2009). Mechanisms for strong adsorption of tetracycline to carbon nanotubes: A comparative study using activated carbon and graphite as adsorbents. *Environmental Science and Technology*, *43*(7), 2322–2327.

Jiang, J., Xiao, F., He, W. M., & Wang, L. (2021). The application of clean production in organic synthesis. *Chinese Chemical Letters*, *32*(5), 1637–1644.

Kabir, M. M., Akter, M. M., Khandaker, S., Gilroyed, B. H., Didar-ul-Alam, M., Hakim, M., & Awual, M. R. (2022). Highly effective agro-waste based functional green adsorbents for toxic chromium (VI) ion removal from wastewater. *Journal of Molecular Liquids*, *347*, 118327.

Kamal, M. H. M. A., Azira, W. M. K. W. K., Kasmawati, M., Haslizaidi, Z., & Saime, W. N. W. (2010). Sequestration of toxic Pb (II) ions by chemically treated rubber (Hevea brasiliensis) leaf powder. *Journal of Environmental Sciences*, *22*(2), 248–256.

Kamm, B., & Kamm, M. (2004). Biorefinery-systems. *Chemical and Biochemical Engineering Quarterly*, *18*(1), 1–7.

Karimi, F., Ayati, A., Tanhaei, B., Sanati, A. L., Afshar, S., Kardan, A., Dabirifar, Z., & Karaman, C. (2022). Removal of metal ions using a new magnetic chitosan nano-bio-adsorbent; A powerful approach in water treatment. *Environmental Research*, *203*, 111753.

Karthikeyan, M., Renganathan, S., & Baskar, G. (2017). Production of biodiesel from waste cooking oil using $MgMoO_4$-supported TiO_2 as a heterogeneous catalyst. *Energy Sources: Part A*, *39*(21), 2053–2059.

Kaur, G., Kumar, H., & Singla, M. (2022). Diverse applications of ionic liquids: A comprehensive review. *Journal of Molecular Liquids*, *351*, 118556.

Kaur, K., Jindal, R., & Saini, D. (2020). Synthesis, optimization and characterization of PVA-co-poly (methacrylic acid) green adsorbents and applications in environmental remediation. *Polymer Bulletin*, *77*(6), 3079–3100.

Kebede, T. G., Dube, S., Mengistie, A. A., Nkambule, T. T., & Nindi, M. M. (2018). Moringa stenopetala bark: A novel green adsorbent for the removal of metal ions from industrial effluents. *Physics and Chemistry of the Earth: Parts A/B/C*, *107*, 45–57.

Khalatbary, M., Sayadi, M. H., Hajiani, M., Nowrouzi, M., & Homaeigohar, S. (2022). Green, sustainable synthesis of γ-Fe_2O_3/MWCNT/Ag nano-composites using the Viscum album leaf extract and waste car tire for removal of sulfamethazine and bacteria from wastewater streams. *Nanomaterials*, *12*(16), 2798.

Khamis Soliman, N., Moustafa, A. F., Aboud, A. A., & Halim, K. S. A. (2019). Effective utilization of Moringa seeds waste as a new green environmental adsorbent for removal of industrial toxic dyes. *Journal of Materials Research and Technology*, *8*(2), 1798–1808.

Khan, S. H., & Iqbal, J. (2016). Recent advances in the role of organic acids in poultry nutrition. *Journal of Applied Animal Research*, *44*(1), 359–369.

Kim, J. H., Kang, Y. J., Kim, K. I., Kim, S. K., & Kim, J. H. (2019). Toxic effects of nitrogenous compounds (ammonia, nitrite, and nitrate) on acute toxicity and anti-oxidant responses of juvenile olive flounder, Paralichthys olivaceus. *Environmental Toxicology and Pharmacology*, *67*, 73–78.

Kirchhoff, M. M. (2005). Promoting sustainability through green chemistry. *Resources, Conservation and Recycling*, *44*(3), 237–243.

Kokel, A., & Török, B. (2018). Sustainable production of fine chemicals and materials using nontoxic renewable sources. *Toxicological Sciences*, *161*(2), 214–224.

Kottappara, R., Palantavida, S., & Vijayan, B. K. (2020). Reduce derivatives. In *Green chemistry and applications*, Sáenz-Galindo, Facio, A., Rodriguez-Herrera, R. (Eds.), (pp. 152–176). CRC Press.

Krylova, A. Y., Kozyukov, E. A., & Lapidus, A. L. (2008). Ethanol and diesel fuel from plant raw materials: A review. *Solid Fuel Chemistry*, *42*(6), 358–364.

Kumar, D., Das, T., Giri, B. S., Rene, E. R., & Verma, B. (2019). Biodiesel production from hybrid non-edible oil using bio-support beads immobilized with lipase from Pseudomonas cepacia. *Fuel*, *255*, 115801.

Kuppusamy, S., Thavamani, P., Megharaj, M., Venkateswarlu, K., Lee, Y. B., & Naidu, R. (2016). Potential of Melaleuca diosmifolia leaf as a low-cost adsorbent for hexavalent chromium removal from contaminated water bodies. *Process Safety and Environmental Protection*, *100*, 173–182.

Kürti, L., Veszelka, S., Bocsik, A., Dung, N. T. K., Ózsvári, B., Puskás, L. G., Kittel, A., Szabó-Révész, P., & Deli, M. A. (2012). The effect of sucrose esters on a culture model of the nasal barrier. *Toxicology in Vitro*, *26*(3), 445–454.

Kyzas, G. Z., & Deliyanni, E. A. (2015). Modified activated carbons from potato peels as green environmental-friendly adsorbents for the treatment of pharmaceutical effluents. *Chemical Engineering Research and Design*, *97*, 135–144.

Kyzas, G. Z., Kostoglou, M., Lazaridis, N. K., Lambropoulou, D. A., & Bikiaris, D. N. (2013). Environmental friendly technology for the removal of pharmaceutical contaminants from wastewaters using modified chitosan adsorbents. *Chemical Engineering Journal*, *222*, 248–258.

Lang, Q., & Wai, C. M. (2001). Supercritical fluid extraction in herbal and natural product studies—A practical review. *Talanta*, *53*(4), 771–782.

Laughton, M. A. (1990). *Renewable energy sources: Watt committee: Report number 22*. CRC Press.

Lebl, R., Murray, T., Adamo, A., Cantillo, D., & Kappe, C. O. (2019). Continuous flow synthesis of methyl Oximino acetoacetate: Accessing greener purification methods with inline liquid–liquid extraction and membrane separation technology. *ACS Sustainable Chemistry and Engineering*, *7*(24), 20088–20096.

Lemos, V. A., Teixeira, L. S. G., Bezerra, M. D. A., Costa, A. C. S., Castro, J. T., Cardoso, L. A. M., de Jesus, D. S., Santos, E. S., Baliza, P. X., & Santos, L. N. (2008). New materials for solid-phase extraction of trace elements. *Applied Spectroscopy Reviews*, *43*(4), 303–334.

Levard, C., Hamdi-Alaoui, K., Baudin, I., Guillon, A., Borschneck, D., Campos, A., Bizi, M., Benoit, F., Chaneac, C., & Labille, J. (2021). Silica-clay nanocomposites for the removal of antibiotics in the water usage cycle. *Environmental Science and Pollution Research*, *28*(6), 7564–7573.

Li, C., Zhou, K., Qin, W., Tian, C., Qi, M., Yan, X., & Han, W. (2019). A review on heavy metals contamination in soil: Effects, sources, and remediation techniques. *Soil and Sediment Contamination: An International Journal*, *28*(4), 380–394.

Li, W., Chai, L., Du, B., Chen, X., & Sun, R. C. (2023). Full-lignin-based adsorbent for removal of Cr(VI) from waste water. *Separation and Purification Technology*, *306*, 122644.

Lindström, U. M. (2002). Stereoselective organic reactions in water. *Chemical Reviews*, *102*(8), 2751–2772.

Liu, H., Jiang, L., Lu, M., Liu, G., Li, T., Xu, X., Li, L., Lin, H., Lv, J., Huang, X., & Xu, D. (2019a). Magnetic solid-phase extraction of pyrethroid pesticides from environmental water samples using deep eutectic solvent-type surfactant modified magnetic zeolitic imidazolate framework-8. *Molecules*, *24*(22), 4038.

Liu, Y., Liu, R., Li, M., Yu, F., & He, C. (2019b). Removal of pharmaceuticals by novel magnetic genipin-crosslinked chitosan/graphene oxide-SO3H composite. *Carbohydrate Polymers*, *220*, 141–148.

Liu, Y., & Liu, Y. J. (2008). Biosorption isotherms, kinetics and thermodynamics. *Separation and Purification Technology*, *61*(3), 229–242.

Liyanage, A. S., Canaday, S., Pittman Jr., C. U., & Mlsna, T. (2020). Rapid remediation of pharmaceuticals from wastewater using magnetic Fe_3O_4/Douglas fir biochar adsorbents. *Chemosphere*, *258*, 127336.

Logar, N. (2011). Chemistry, green chemistry, and the instrumental valuation of sustainability. *Minerva*, *49*(1), 113–136.

Lokesh, K., West, C., Kuylenstierna, J. C., Fan, J., Budarin, V., Priecel, P., Lopez-Sanchez, J. A., & Clark, J. H. (2019). Economic and agronomic impact assessment of wheat straw based alkyl polyglucoside produced using green chemical approaches. *Journal of Cleaner Production*, *209*, 283–296.

Lokesh, K., West, C., Kuylenstierna, J., Fan, J., Budarin, V., Priecel, P., Lopez-Sanchez, J. A., & Clark, J. (2017). Environmental impact assessment of wheat straw based alkyl polyglucosides produced using novel chemical approaches. *Green Chemistry*, *19*(18), 4380–4395.

Lonappan, L., Brar, S. K., Das, R. K., Verma, M., & Surampalli, R. Y. (2016). Diclofenac and its transformation products: Environmental occurrence and toxicity-a review. *Environment International*, *96*, 127–138.

Lonappan, L., Liu, Y., Rouissi, T., Brar, S. K., & Surampalli, R. Y. (2020). Development of biochar-based green functional materials using organic acids for environmental applications. *Journal of Cleaner Production*, *244*, 118841.

Lu, C., Chiu, H., & Liu, C. (2006). Removal of zinc (II) from aqueous solution by purified carbon nanotubes: kinetics and equilibrium studies. *Industrial and Engineering Chemistry Research*, *45*(8), 2850–2855.

Lu, Y., Priyantha, N., & Lim, L. B. (2020). *Ipomoea aquatica* roots as environmentally friendly and green adsorbent for efficient removal of Auramine O dye. *Surfaces and Interfaces*, *20*, 100543.

Luo, B., Ye, D., & Wang, L. (2017). Recent progress on integrated energy conversion and storage systems. *Advanced Science*, *4*(9), 1700104.

Luque, R., Herrero-Davila, L., Campelo, J. M., Clark, J. H., Hidalgo, J. M., Luna, D., Marinas, J. M., & Romero, A. A. (2008). Biofuels: A technological perspective. *Energy and Environmental Science*, *1*(5), 542–564.

Ma, J., Wang, Y., Liu, G., Xu, N., & Wang, X. (2020). A pH-stable Ag (I) multifunctional luminescent sensor for the efficient detection of organic solvents, organochlorine pesticides and heavy metal ions. *RSC Advances*, *10*(73), 44712–44718.

Macedo, J. C. A., Gontijo, E. S. J., Herrera, S. G., Rangel, E. C., Komatsu, D., Landers, R., & Rosa, A. H. (2021). Organosulphur-modified biochar: An effective green adsorbent for removing metal species in aquatic systems. *Surfaces and Interfaces*, *22*, 100822.

Mahamadi, C. (2011). Water hyacinth as a biosorbent: A review. *African Journal of Environmental Science and Technology*, *5*(13), 1137–1145.

Mahmoodi, N. M., Taghizadeh, M., & Taghizadeh, A. (2019). Activated carbon/metal-organic framework composite as a bio-based novel green adsorbent: Preparation and mathematical pollutant removal modeling. *Journal of Molecular Liquids*, *277*, 310–322.

Maity, S. K. (2015). Opportunities, recent trends and challenges of integrated biorefinery: Part I. *Renewable and Sustainable Energy Reviews*, *43*, 1427–1445.

Mallakpour, S., & Soltanian, S. (2016). Surface functionalization of carbon nanotubes: Fabrication and applications. *RSC Advances*, *6*(111), 109916–109935.

Manko, D., & Zdziennicka, A. (2015). Sugar-based surfactants as alternative to synthetic ones. *Ann. UMCS Chem.*, *70*, 161–168.

Mathers, R. T., McMahon, K. C., Damodaran, K., Retarides, C. J., & Kelley, D. J. (2006). Ring-opening metathesis polymerizations in D-limonene: A renewable polymerization solvent and chain transfer agent for the synthesis of alkene macromonomers. *Macromolecules*, *39*(26), 8982–8986.

Meshkat, S. S., Nezhad, M. N., & Bazmi, M. R. (2019). Investigation of Carmine dye removal by green chitin nanowhiskers adsorbent. *Emerging Science Journal*, *3*(3), 187–194.

Mikkola, J. P., Sklavounos, E., King, A. W., & Virtanen, P. (2015). The biorefinery and green chemistry. In *Ionic Liquids in the Biorefinery Concept: Challenges and Perspectives*, Bogel-Lukasik, R. (Ed.), (pp. 1–37). The Royal Society of Chemistry Publication.

Mitra, T., & Das, S. K. (2019). Cr (VI) removal from aqueous solution using Psidium guajava leaves as green adsorbent: Column studies. *Applied Water Science*, *9*(7), 1–8.

Moldes, A. B., Rodríguez-López, L., Rincón-Fontán, M., López-Prieto, A., Vecino, X., & Cruz, J. M. (2021). Synthetic and bio-derived surfactants versus microbial biosurfactants in the cosmetic industry: An overview. *International Journal of Molecular Sciences*, *22*(5), 2371.

Mondal, D. K., Nandi, B. K., & Purkait, M. K. (2013). Removal of mercury (II) from aqueous solution using bamboo leaf powder: equilibrium, thermodynamic and kinetic studies. *Journal of Environmental Chemical Engineering*, *1*(4), 891–898.

Mondal, S., Patel, S., & Majumder, S. K. (2020). Bio-extract assisted in-situ green synthesis of Ag-RGO nanocomposite film for enhanced naproxen removal. *Korean Journal of Chemical Engineering*, *37*(2), 274–289.

Mousavi, S. H., Samakchi, S., Kianfar, M., Zahmatkesh, M., & Zohourtalab, A. (2022). Raw sunflower seed shell particles as a green adsorbent for the adsorption of synthetic organic dyes from wastewater. *Environmental Progress and Sustainable Energy*, *42*(2), e14017.

Mukherjee, P. (2021). Green chemistry–A novel approach towards sustainability. *Journal of the Chilean Chemical Society*, *66*(1), 5075–5080.

Munir, R., Ali, K., Naqvi, S. A. Z., Maqsood, M. A., Bashir, M. Z., & Noreen, S. (2023). Biosynthesis of Leucaena leucocephala leaf mediated ZnO, CuO, MnO_2, and MgO based nano-adsorbents for Reactive Golden Yellow-145 (RY-145) and Direct Red-31 (DR-31) dye removal from textile wastewater to reuse in agricultural purpose. *Separation and Purification Technology*, *306*, 122527.

Musarurwa, H., & Tavengwa, N. T. (2021). Extraction and electrochemical sensing of pesticides in food and environmental samples by use of polydopamine-based materials. *Chemosphere*, *266*, 129222.

Naghdi, M., Taheran, M., Brar, S. K., Kermanshahi-Pour, A., Verma, M., & Surampalli, R. Y. (2017). Immobilized laccase on oxygen functionalized nanobiochars through mineral acids treatment for removal of carbamazepine. *Science of the Total Environment*, *584*, 393–401.

Nasiri, M., Ahmadzadeh, H., & Amiri, A. (2022). Magnetic solid-phase extraction of organophosphorus pesticides from apple juice and environmental water samples using magnetic graphene oxide coated with poly (2-aminoterephthalic acid-co-aniline) nanocomposite as a sorbent. *Journal of Separation Science*, *45*(13), 2301–2309.

Naughton, M. J., & Drago, R. S. (1995). Supported homogeneous film catalysts. *Journal of Catalysis*, *155*(2), 383–389.

Naushad, M., Ahamad, T., AlOthman, Z. A., & Ala'a, H. (2019). Green and eco-friendly nanocomposite for the removal of toxic Hg (II) metal ion from aqueous environment: Adsorption kinetics & isotherm modelling. *Journal of Molecular Liquids*, *279*, 1–8.

Naveenkumar, R., & Baskar, G. (2020). Optimization and techno-economic analysis of biodiesel production from Calophyllum inophyllum oil using heterogeneous nanocatalyst. *Bioresource Technology*, *315*, 123852.

Naveenkumar, R., & Baskar, G. (2021). Process optimization, green chemistry balance and technoeconomic analysis of biodiesel production from castor oil using heterogeneous nanocatalyst. *Bioresource Technology*, *320*(A), 124347.

Nazraz, M., Yamini, Y., & Asiabi, H. (2019). Chitosan-based sorbent for efficient removal and extraction of ciprofloxacin and norfloxacin from aqueous solutions. *Microchimica Acta*, *186*(7), 1–9.

Ncibi, M. C., & Sillanpää, M. (2017). Optimizing the removal of pharmaceutical drugs carbamazepine and dorzolamide from aqueous solutions using mesoporous activated carbons and multi-walled carbon nanotubes. *Journal of Molecular Liquids*, *238*, 379–388.

Neta, N. S., Teixeira, J. A., & Rodrigues, L. R. (2015). Sugar ester surfactants: Enzymatic synthesis and applications in food industry. *Critical Reviews in Food Science and Nutrition*, *55*(5), 595–610.

Nguyen, H. T., Ngwabebhoh, F. A., Saha, N., Saha, T., & Saha, P. (2022). Gellan gum/bacterial cellulose hydrogel crosslinked with citric acid as an eco-friendly green adsorbent for safranin and crystal violet dye removal. *International Journal of Biological Macromolecules*, *222*(A), 77–89.

Nyaga, M. N., Nyagah, D. M., & Njagi, A. (2020). Pharmaceutical waste: Overview, management, and impact of improper disposal. *Journal of PeerScientist 3*(2), e1000028.

O'Flynn, D., Lawler, J., Yusuf, A., Parle-McDermott, A., Harold, D., McCloughlin, T., Holland, L., Regan, F., & White, B. (2021). A review of pharmaceutical occurrence and pathways in the aquatic environment in the context of a changing climate and the COVID-19 pandemic. *Analytical Methods*, *13*(5), 575–594.

Okenicová, L., Žemberyová, M., & Procházková, S. (2016). Biosorbents for solid-phase extraction of toxic elements in waters. *Environmental Chemistry Letters*, *14*(1), 67–77.

Omodara, L., Saavalainen, P., Pitkäaho, S., Pongrácz, E., & Keiski, R. L. (2023). Sustainability assessment of products-case study of wind turbine generator types. *Environmental Impact Assessment Review*, *98*, 106943.

Otache, M. A., Duru, R. U., Ozioma, A., & Abayeh, J. O. (2022). Catalytic methods for the synthesis of sugar esters. *Catalysis in Industry*, *14*(1), 115–130.

Otunola, B. O., & Ololade, O. O. (2020). A review on the application of clay minerals as heavy metal adsorbents for remediation purposes. *Environmental Technology and Innovation*, *18*, 100692.

Outili, N., Kerras, H., Nekkab, C., Merouani, R., & Meniai, A. H. (2020). Biodiesel production optimization from waste cooking oil using green chemistry metrics. *Renewable Energy*, *145*, 2575–2586.

Pan, S. Z., Jin, C. Z., Yang, X. A., & Zhang, W. B. (2020). Ultrasound enhanced solid-phase extraction of ultra-trace arsenic on Fe3O4@ AuNPs magnetic particles. *Talanta*, *209*, 120553.

Pârvulescu, V. I., & Hardacre, C. (2007). Catalysis in ionic liquids. *Chemical Reviews*, *107*(6), 2615–2665.

Patel, M. (2003). Surfactants based on renewable raw materials: Carbon dioxide reduction potential and policies and measures for the European Union. *Journal of Industrial Ecology*, *7*(3–4), 47–62.

Pena-Pereira, F., Kloskowski, A., & Namieśnik, J. (2015). Perspectives on the replacement of harmful organic solvents in analytical methodologies: A framework toward the implementation of a generation of eco-friendly alternatives. *Green Chemistry*, *17*(7), 3687–3705.

Perlatti, B., Forim, M. R., & Zuin, V. G. (2014). Green chemistry, sustainable agriculture and processing systems: A Brazilian overview. *Chemical and Biological Technologies in Agriculture*, *1*(1), 1–9.

Petridis, N. P., Sakkas, V. A., & Albanis, T. A. (2014). Chemometric optimization of dispersive suspended microextraction followed by gas chromatography–mass spectrometry for the determination of polycyclic aromatic hydrocarbons in natural waters. *Journal of Chromatography. Part A*, *1355*, 46–52.

Plechkova, N. V., & Seddon, K. R. (2008). Applications of ionic liquids in the chemical industry. *Chemical Society Reviews*, *37*(1), 123–150.

Pletnev, M. Y. (2001). Chemistry of surfactants. In *Studies in interface science*, Fainerman, V. B., Möbius, D., & Miller, R. (Eds.), (Vol. 13, pp. 1–97). Elsevier.

Poliakoff, M., Fitzpatrick, J. M., Farren, T. R., & Anastas, P. T. (2002). Green chemistry: Science and politics of change. *Science*, *297*(5582), 807–810.

Poliakoff, M., George, M. W., Howdle, S. M., Bagratashvili, V. N., Han, B. X., & Yan, H. K. (1999). Supercritical fluids: Clean solvents for green chemistry. *Chinese Journal of Chemistry*, *17*(3), 212–222.

Pollet, P., Davey, E. A., Ureña-Benavides, E. E., Eckert, C. A., & Liotta, C. L. (2014). Solvents for sustainable chemical processes. *Green Chemistry*, *16*(3), 1034–1055.

Prajapati, A. K., Das, S., & Mondal, M. K. (2020). Exhaustive studies on toxic Cr (VI) removal mechanism from aqueous solution using activated carbon of Aloe vera waste leaves. *Journal of Molecular Liquids*, *307*, 112956.

Prauchner, M. J., Sapag, K., & Rodríguez-Reinoso, F. (2016). Tailoring biomass-based activated carbon for CH_4 storage by combining chemical activation with H_3PO_4 or $ZnCl_2$ and physical activation with CO_2. *Carbon*, *110*, 138–147.

Ragheb, E., Shamsipur, M., Jalali, F., & Mousavi, F. (2022). Modified magnetic-metal organic framework as a green and efficient adsorbent for removal of heavy metals. *Journal of Environmental Chemical Engineering*, *10*(2), 107297.

Raj, A., Chowdhury, A., & Ali, S. W. (2022). Green chemistry: Its opportunities and challenges in colouration and chemical finishing of textiles. *Sustainable Chemistry and Pharmacy*, *27*, 100689.

Rao, G. P., Lu, C., & Su, F. (2007). Sorption of divalent metal ions from aqueous solution by carbon nanotubes: A review. *Separation and Purification Technology*, *58*(1), 224–231.

Rass-Hansen, J., Falsig, H., Jørgensen, B., & Christensen, C. H. (2007). Bioethanol: Fuel or feedstock? *Journal of Chemical Technology & Biotechnology: International Research in Process, Environmental & Clean Technology*, *82*(4), 329–333.

Rathnakumar, S., Sheeja, R. Y., & Murugesan, T. (2009). Removal of copper (II) from aqueous solutions using teak (Tectona grandis Lf) leaves. *World Academy of Science, Engineering and Technology*, *56*, 880–884.

Rayner, C. M. (2007). The potential of carbon dioxide in synthetic organic chemistry. *Organic Process Research and Development*, *11*(1), 121–132.

Reyes-Calderón, A., Pérez-Uribe, S., Ramos-Delgado, A. G., Ramalingam, S., Oza, G., Parra-Saldívar, R., Ramirez-Mendoza, R. A., Iqbal, H. M. N., & Sharma, A. (2022). Analytical and regulatory considerations to mitigate highly hazardous toxins from environmental matrices. *Journal of Hazardous Materials*, *423*(A), 127031.

Rutkowska, J., & Stolyhwo, A. (2009). Application of carbon dioxide in subcritical state (LCO2) for extraction/fractionation of carotenoids from red paprika. *Food Chemistry*, *115*(2), 745–752.

Ryan, A. A., & Senge, M. O. (2015). How green is green chemistry? Chlorophylls as a bioresource from biorefineries and their commercial potential in medicine and photovoltaics. *Photochemical and Photobiological Sciences*, *14*(4), 638–660.

Sakaki, K., Aoyama, A., Nakane, T., Ikegami, T., Negishi, H., Watanabe, K., & Yanagishita, H. (2006). Enzymatic synthesis of sugar esters in organic solvent coupled with pervaporation. *Desalination*, *193*(1–3), 260–266.

Saleh, H. E. D. M., & Koller, M. (2018). Introductory chapter: Principles of green chemistry. In *Green chemistry*, Saleh, H. E. D. M., & Koller, M. (Eds.), (pp. 3–15) IntechOpen.

Samadi, S., Sereshti, H., & Assadi, Y. (2012). Ultra-preconcentration and determination of thirteen organophosphorus pesticides in water samples using solid-phase extraction followed by dispersive liquid–liquid microextraction and gas chromatography with flame photometric detection. *Journal of Chromatography: Part A*, *1219*, 61–65.

Sanghi, R. (2000). Better living through 'sustainable green chemistry'. *Current Science*, *79*(12), 1662–1665.

Sans, V., & Cronin, L. (2016). Towards dial-a-molecule by integrating continuous flow, analytics and self-optimisation. *Chemical Society Reviews*, *45*(8), 2032–2043.

Sarkar, N., Ghosh, S. K., Bannerjee, S., & Aikat, K. (2012). Bioethanol production from agricultural wastes: An overview. *Renewable Energy*, *37*(1), 19–27.

Schulte, P. A., McKernan, L. T., Heidel, D. S., Okun, A. H., Dotson, G. S., Lentz, T. J., Geraci, C. L., Heckel, P. E., & Branche, C. M. (2013). Occupational safety and health, green chemistry, and sustainability: A review of areas of convergence. *Environmental Health*, *12*(1), 1–9.

Schwarzenbach, R. P., Escher, B. I., Fenner, K., Hofstetter, T. B., Johnson, C. A., Von Gunten, U., & Wehrli, B. (2006). The challenge of micropollutants in aquatic systems. *Science*, *313*(5790), 1072–1077.

Sculley, J., Yuan, D., & Zhou, H. C. (2011). The current status of hydrogen storage in metal–organic frameworks—updated. *Energy & Environmental Science*, *4*(8), 2721–2735.

Sebayang, A. H., Masjuki, H. H., Ong, H. C., Dharma, S., Silitonga, A. S., Mahlia, T. M. I., & Aditiya, H. B. (2016). A perspective on bioethanol production from biomass as alternative fuel for spark ignition engine. *RSC Advances*, *6*(18), 14964–14992.

Sharma, P., Prakash, J., & Kaushal, R. (2022). An insight into the green synthesis of SiO_2 nanostructures as a novel adsorbent for removal of toxic water pollutants. *Environmental Research*, *212*(C), 113328.

Sharma, S. K., Chaudhary, A., & Singh, R. V. (2008). Gray chemistry verses green chemistry: Challenges and opportunities. *Rasayan Journal of Chemistry*, *1*, 68–92.

Sharma, Y. C., Srivastava, V., Singh, V. K., Kaul, S. N., & Weng, C. H. (2009). Nano-adsorbents for the removal of metallic pollutants from water and wastewater. *Environmental Technology*, *30*(6), 583–609.

Shi, J., Fang, Z., Zhao, Z., Sun, T., & Liang, Z. (2016). Comparative study on Pb (II), Cu (II), and Co (II) ions adsorption from aqueous solutions by arborvitae leaves. *Desalination and Water Treatment*, *57*(10), 4732–4739.

Shi, X., Cheng, C., Peng, F., Hou, W., Lin, X., & Wang, X. (2022). Adsorption properties of graphene materials for pesticides: Structure effect. *Journal of Molecular Liquids*, *364*, 119967.

Shonnard, D., Lindner, A., Nguyen, N., Ramachandran, P. A., Fichana, D., Hesketh, R., Stewart Slater, C., & Engler, R. (2012). Green engineering: Integration of green chemistry, pollution prevention, and risk-based considerations. In *Handbook of industrial chemistry and biotechnology,* Kent, J. (Eds.), (pp. 155–199). Springer.

Simon, M. O., & Li, C. J. (2012). Green chemistry oriented organic synthesis in water. *Chemical Society Reviews*, *41*(4), 1415–1427.

Sinha, E., Michalak, A. M., & Balaji, V. (2017). Eutrophication will increase during the 21st century as a result of precipitation changes. *Science*, *357*(6349), 405–408.

Song, X., Zhang, R., Xie, T., Wang, S., & Cao, J. (2019). Deep eutectic solvent micro-functionalized graphene assisted dispersive micro solid-phase extraction of pyrethroid insecticides in natural products. *Frontiers in Chemistry*, *7*, 594.

Spaolonzi, M. P., Duarte, E. D., Oliveira, M. G., Costa, H. P., Ribeiro, M. C., Silva, T. L., Silva, M. G. C., & Vieira, M. G. (2022). Green-functionalized carbon nanotubes as adsorbents for the removal of emerging contaminants from aqueous media. *Journal of Cleaner Production*, *373*, 133961.

Spear, S. K., Griffin, S. T., Granger, K. S., Huddleston, J. G., & Rogers, R. D. (2007). Renewable plant-based soybean oil methyl esters as alternatives to organic solvents. *Green Chemistry*, *9*(9), 1008–1015.

Stubbs, S., Yousaf, S., & Khan, I. (2022). A review on the synthesis of bio-based surfactants using green chemistry principles. *DARU: Journal of Pharmaceutical Sciences, 30*(2), 1–20

Suri, A., Khandegar, V., & Kaur, P. J. (2021). Ofloxacin exclusion using novel HRP immobilized chitosan cross-link with graphene-oxide nanocomposite. *Groundwater for Sustainable Development, 12*, 100515.

Szűts, A., Budai-Szűcs, M., Erős, I., Otomo, N., & Szabó-Révész, P. (2010). Study of gel-forming properties of sucrose esters for thermosensitive drug delivery systems. *International Journal of Pharmaceutics, 383*(1–2), 132–137.

Szűts, A., & Szabó-Révész, P. (2012). Sucrose esters as natural surfactants in drug delivery systems—A mini-review. *International Journal of Pharmaceutics, 433*(1–2), 1–9.

Tan, C. W., Tan, K. H., Ong, Y. T., Mohamed, A. R., Zein, S. H. S., & Tan, S. H. (2012). Energy and environmental applications of carbon nanotubes. *Environmental Chemistry Letters, 10*(3), 265–273.

Telkapalliwar, N. G., & Shivankar, V. M. (2018). Adsorption of zinc onto microwave assisted carbonized Acacia nilotica bark. *Materials Today: Proceedings, 5*(10), 22694–22705.

Teng, D., Zhang, B., Xu, G., Wang, B., Mao, K., Wang, J., Sun, J., Feng, X., Yang, Z., & Zhang, H. (2020). Efficient removal of Cd (II) from aqueous solution by pinecone biochar: Sorption performance and governing mechanisms. *Environmental Pollution, 265*(A), 115001.

Tian, C., Wu, Z., He, M., Chen, B., & Hu, B. (2022). Amino functionalized magnetic covalent organic framework for magnetic solid-phase extraction of sulfonylurea herbicides in environmental samples from tobacco land. *Journal of Separation Science, 45*(10), 1746–1756.

Tobiszewski, M., Marć, M., Gałuszka, A., & Namieśnik, J. (2015). Green chemistry metrics with special reference to green analytical chemistry. *Molecules, 20*(6), 10928–10946.

Todosijević, M. N., Cekić, N. D., Savić, M. M., Gašperlin, M., Ranđelović, D. V., & Savić, S. D. (2014). Sucrose ester-based biocompatible microemulsions as vehicles for aceclofenac as a model drug: Formulation approach using D-optimal mixture design. *Colloid and Polymer Science, 292*(12), 3061–3076.

Trivedi, P., & Axe, L. (2000). Modeling Cd and Zn sorption to hydrous metal oxides. *Environmental Science and Technology, 34*(11), 2215–2223.

Trost, B. M. (1991). The atom economy—A search for synthetic efficiency. *Science, 254*(5037), 1471–1477.

Tundo, P., Anastas, P., Black, D. S., Breen, J., Collins, T. J., Memoli, S., Miyamoto, J., Polyakoff, M., & Tumas, W. (2000). Synthetic pathways and processes in green chemistry. Introductory overview. *Pure and Applied Chemistry, 72*(7), 1207–1228.

Türker, A. R. (2007). New sorbents for solid-phase extraction for metal enrichment. *CLEAN–Soil, Air, Water, 35*(6), 548–557.

Tüysüz, M., Köse, K., Aksüt, D., Uzun, L., Evci, M., Köse, D. A., & Youngblood, J. P. (2022). Removal of atrazine using polymeric cryogels modified with cellulose nanomaterials. *Water, Air, and Soil Pollution, 233*(11), 1–13.

van Rantwijk, F., & Sheldon, R. A. (2007). Biocatalysis in ionic liquids. *Chemical Reviews, 107*(6), 2757–2785.

Van Tran, T., Nguyen, D. T. C., Kumar, P. S., Din, A. T. M., Qazaq, A. S., & Vo, D. V. N. (2022). Green synthesis of Mn_3O_4 nanoparticles using *Costus woodsonii* flowers extract for effective removal of malachite green dye. *Environmental Research, 214*(2), 113925.

Vargas, J. A. M., Ortega, J. O., Dos Santos, M. B. C., Metzker, G., Gomes, E., & Boscolo, M. (2020). A new synthetic methodology for pyridinic sucrose esters and their antibacterial effects against Gram-positive and Gram-negative strains. *Carbohydrate Research, 489,* 107957.

Vijayaraghavan, K., & Yun, Y. S. (2008). Bacterial biosorbents and biosorption. *Biotechnology Advances, 26*(3), 266–291.

Villabona-Ortíz, Á., Tejada-Tovar, C., & Gonzalez-Delgado, Á. D. (2021). Adsorption of CD2+ ions from aqueous solution using biomasses of Theobroma cacao, Zea mays, Manihot esculenta, Dioscorea rotundata and Elaeis guineensis. *Applied Sciences, 11*(6), 2657.

Villaseñor, M., & Ríos, Á. (2018). Nanomaterials for water cleaning and desalination, energy production, disinfection, agriculture and green chemistry. *Environmental Chemistry Letters, 16*(1), 11–34.

Vishwakarma, S. (2014). Ionic liquids-designer solvents for green chemistry. *International Journal of Basic Sciences and Applied Computing (IJBSAC), 1*(1), 1–4.

Wan Ibrahim, W. A., Abd Ali, L. I., Sulaiman, A., Sanagi, M. M., & Aboul-Enein, H. Y. (2014). Application of solid-phase extraction for trace elements in environmental and biological samples: A review. *Critical Reviews in Analytical Chemistry, 44*(3), 233–254.

Wang, C., & Wang, H. (2018). Carboxyl functionalized *Cinnamomum camphora* for removal of heavy metals from synthetic wastewater-contribution to sustainability in agroforestry. *Journal of Cleaner Production, 184,* 921–928.

Wang, T., Zhang, R., Gong, Z., Su, P., & Yang, Y. (2020). Poly (ionic liquids) functionalized magnetic nanoparticles as efficient adsorbent for determination of pyrethroids from environmental water samples by GC-MS. *ChemistrySelect, 5*(1), 91–96.

Wardencki, W., Curyło, J., & Namieśnik, J. (2005). Green chemistry–Current and future issues. *Polish Journal of Environmental Studies, 14*(4), 389–395.

Weber Jr., W. J., McGinley, P. M., & Katz, L. E. (1991). Sorption phenomena in subsurface systems: Concepts, models and effects on contaminant fate and transport. *Water Research, 25*(5), 499–528.

Welton, T. (1999). Room-temperature ionic liquids. Solvents for synthesis and catalysis. *Chemical Reviews, 99*(8), 2071–2084.

Welton, T. (2015). Solvents and sustainable chemistry. *Proceedings of the Royal Society A: Mathematical, Physical and Engineering Sciences, 471*(2183), 20150502.

Whiteker, G. T. (2019). Applications of the 12 principles of green chemistry in the crop protection industry. *Organic Process Research and Development, 23*(10), 2109–2121.

Wijewardane, S. (2009). Potential applicability of CNT and CNT/composites to implement ASEC concept: A review article. *Solar Energy, 83*(8), 1379–1389.

Wilkinson, S. L. (1997). Green is practical, even profitable. *Chemical & Engineering News, 75*(31), 35–35.

Xiao, Y. M., Wu, Q., Cai, Y., & Lin, X. F. (2005). Ultrasound-accelerated enzymatic synthesis of sugar esters in nonaqueous solvents. *Carbohydrate Research, 340*(13), 2097–2103.

Yadav, G. D. (2006). Green chemistry in India. *Clean Technologies and Environmental Policy, 8*(4), 219–223.

Yadav, S., Tam, J., & Singh, C. V. (2015). A first principles study of hydrogen storage on lithium decorated two dimensional carbon allotropes. *International journal of hydrogen energy, 40*(18), 6128–6136.

Yang, K. U. N., Wu, W., Jing, Q., & Zhu, L. (2008). Aqueous adsorption of aniline, phenol, and their substitutes by multi-walled carbon nanotubes. *Environmental Science and Technology, 42*(21), 7931–7936.

Yang, K., & Xing, B. (2010). Adsorption of organic compounds by carbon nanomaterials in aqueous phase: Polanyi theory and its application. *Chemical Reviews*, *110*(10), 5989–6008.

Yang, L., Zhan, Y., Gong, Y., Ren, E., Lan, J., Guo, R., Yan, B., Chen, S., & Lin, S. (2021). Development of eco-friendly CO2-responsive cellulose nanofibril aerogels as "green" adsorbents for anionic dyes removal. *Journal of Hazardous Materials*, *405*, 124194.

Ye, R., Pyo, S. H., & Hayes, D. G. (2010). Lipase-catalyzed synthesis of saccharide–fatty acid esters using suspensions of saccharide crystals in solvent-free media. *Journal of the American Oil Chemists' Society*, *87*(3), 281–293.

Yeganeh, M. S., Kazemizadeh, A. R., Ramazani, A., Eskandari, P., & Angourani, H. R. (2020). Plant-mediated synthesis of $Cu_{0.5}Zn_{0.5}Fe_2O_4$ nanoparticles using *Minidium laevigatum* and their applications as an adsorbent for removal of reactive blue 222 dye. *Materials Research Express*, *6*(12), 1250f4.

Yogeshwaran, V., & Priya, A. K. (2021). Adsorption of lead ion concentration from the aqueous solution using tobacco leaves. *Materials Today: Proceedings*, *37*, 489–496.

Yoo, I. S., Park, S. J., & Yoon, H. H. (2007). Enzymatic synthesis of sugar fatty acid esters. *Journal of Industrial and Engineering Chemistry*, *13*(1), 1–6.

Yousefi, S. M., Shemirani, F., & Ghorbanian, S. A. (2017). Deep eutectic solvent magnetic bucky gels in developing dispersive solid phase extraction: Application for ultra trace analysis of organochlorine pesticides by GC-micro ECD using a large-volume injection technique. *Talanta*, *168*, 73–81.

Yu, J. X., Feng, L. Y., Cai, X. L., Wang, L. Y., & Chi, R. A. (2015). Adsorption of Cu^{2+}, CD^{2+} and Zn^{2+} in a modified leaf fixed-bed column: Competition and kinetics. *Environmental Earth Sciences*, *73*(4), 1789–1798.

Yue, C. S., Chong, K. H., Eng, C. C., & Loh, L. S. (2016). Utilization of infused tea leaves (Camellia sinensis) for the removal of Pb2+, Fe2+ and Cd2+ ions from aqueous solution: Equilibrium and kinetic studies. *Journal of Water Resource and Protection*, *8*(5), 568.

Zeković, Z., Lepojeviíc, Ž., & Vujić, D. (2000). Supercritical extraction of thyme (Thymus vulgaris L.). *Chromatographia*, *51*(3), 175–179.

Zeng, X., & Li, J. (2021). Emerging anthropogenic circularity science: Principles, practices, and challenges. *Iscience*, *24*(3), 102237.

Zhang, F., Gu, W., Xu, P., Tang, S., Xie, K., Huang, X., & Huang, Q. (2011). Effects of alkyl polyglycoside (APG) on composting of agricultural wastes. *Waste Management*, *31*(6), 1333–1338.

Zhang, H., Lai, H., Wu, X., Li, G., & Hu, Y. (2020). $CoFe_2O_4$@ HNTs/AuNPs substrate for rapid magnetic solid-phase extraction and efficient SERS detection of complex samples all-in-one. *Analytical Chemistry*, *92*(6), 4607–4613.

Zhang, M., Yang, J., Geng, X., Li, Y., Zha, Z., Cui, S., & Yang, J. (2019). Magnetic adsorbent based on mesoporous silica nanoparticles for magnetic solid phase extraction of pyrethroid pesticides in water samples. *Journal of Chromatography: Part A*, *1598*, 20–29.

Zhang, Q., Suresh, L., Liang, Q., Zhang, Y., Yang, L., Paul, N., & Tan, S. C. (2021). Emerging technologies for green energy conversion and storage. *Advanced Sustainable Systems*, *5*(3), 2000152.

Zhang, X., Gao, X., Huo, P., & Yan, Y. (2012). Selective adsorption of micro ciprofloxacin by molecularly imprinted functionalized polymers appended onto ZnS. *Environmental Technology*, *33*(17), 2019–2025.

Zhao, J., Meng, Z., Zhao, Z., & Zhao, L. (2020). Ultrasound-assisted deep eutectic solvent as green and efficient media combined with functionalized magnetic multi-walled carbon nanotubes as solid-phase extraction to determine pesticide residues in food products. *Food Chemistry*, *310*, 125863.

Zhao, L., Zhong, S., Fang, K., Qian, Z., & Chen, J. (2012). Determination of cadmium (II), cobalt (II), nickel (II), lead (II), zinc (II), and copper (II) in water samples using dual-cloud point extraction and inductively coupled plasma emission spectrometry. *Journal of Hazardous Materials*, *239*, 206–212.

Zhou, S., Yin, J., Ma, Q., Baihetiyaer, B., Sun, J., Zhang, Y., Jiang, Y., Wang J., & Yin, X. (2022). Montmorillonite-reduced graphene oxide composite aerogel (M-rGO): A green adsorbent for the dynamic removal of cadmium and methylene blue from wastewater. *Separation and Purification Technology*, 296, 121416.

Zuorro, A., & Lavecchia, R. (2010). Adsorption of Pb (II) on spent leaves of green and black tea. *American Journal of Applied Sciences*, *7*(2), 153.

5 Catalytic Methods on Wastewater Remediation and Reuse for Circular Economy

Burcu Palas, Süheyda Atalay, and Gülin Ersöz

CONTENTS

DOI: 10.1201/9781003301769-5

5.1 INTRODUCTION

The growing population has led to an increase in the requirement for goods, which resulted in rapid industrialization. The increase in industrial activities increased the generation of industrial wastes. Industrial wastes cause significant environmental problems due to air, water, and soil pollution. The amount and characteristics of wastewaters varied depending on the type of sector. The wastewaters may contain non-biodegradable pollutants including heavy metals, pesticides, and plastics as well as biodegradable materials comprising paper, leather, and wool. Industrial wastewaters can be reactive, carcinogenic–mutagenic, or toxic (Ahmed et al., 2021).

Wastewaters are mainly classified as stormwater runoff, domestic, municipal, agricultural, and industrial wastewaters, and must be treated prior to their discharge into the environment to avoid any harm or risk that they may have on the environment and human health. The major target of wastewater treatment is to prevent environmental pollution and to protect human health. Another important aim of wastewater treatment is to preserve water sources, which are scarce in many regions of the world, by recycling wastewater for reuse in various applications including irrigation. In addition, value-added chemicals present in wastewaters can be valorized (Edokpayi et al., 2017; Samer, 2015).

As the level of freshwater required increases, wastewater reclamation and reuse are in greater demand. Global water consumption for agriculture is around 92% of the total water consumption. Approximately 70% of the fresh water coming from rivers and groundwater sources is used for irrigation purposes. Generally, treated wastewaters are not used as potable water and are utilized in agriculture, land irrigation, and groundwater recharge. Additionally, treated wastewaters can be used for vehicle washing, toilet flushes, firefighting, building construction activities, and cooling in thermal power plants. The worldwide reuse of treated wastewater for agricultural purposes changes between 1.5% and 6.6% (Ofori et al., 2021). The ratio of the reused wastewater is low and need to be increased.

Wastewater treatment and reclamation techniques are based on individual or hybrid physical, chemical, and biological processes. Physical treatment processes are used to remove solids from wastewaters. Chemical processes that are often integrated with physical processes are applied to remove complex contaminants. Biological methods use microorganisms for the removal and degradation of pollutants. The proper characterization of wastewaters is crucial to determine appropriate treatment technologies that will eliminate the pollutants (Yenkie et al., 2021).

Since conventional wastewater treatment techniques are generally ineffective for the removal of recalcitrant organic pollutants in wastewaters, the use of innovative and cost-efficient technologies is significant to increase the reusability of the treated wastewaters. Advanced oxidation processes based on the generation of reactive species such as hydroxyl radicals have been recently used for the removal of contaminants from wastewaters.

The most common advanced oxidation processes include Fenton oxidation, ozonation, electrochemical oxidation, sonolysis, photocatalysis, supercritical water oxidation, catalytic wet oxidation, and various combinations of advanced oxidation constituents like H_2O_2, O_2, O_3, and light sources.

Complete detoxification and mineralization can be achieved by the application of advanced oxidation methods. These processes can convert pollutants into CO_2, H_2O, and mineral acids. Significant amounts of hydroxyl radicals, which are very effective, nonselective oxidizing agents reacting much more rapidly than alternative oxidants, are formed in the advanced oxidation reactions (Kumar and Bansal, 2013).

The major advantages of the advanced oxidation processes are full mineralization of pollutants, harmless end-products, degradation of a wide variety of organic molecules non-selectively, employment in the degradation of refractory compounds resistant to conventional methods, usage in treatment of highly toxic wastewaters, and integration with other treatment processes.

In addition to the chemical wastewater treatment methods, advanced biological wastewater treatment technologies have been effectively applied for the reclamation of wastewaters. Microbial fuel cell is one of the advanced methods for treating wastewater and simultaneously producing energy. It is a bioelectrochemical method converting the chemical energy of organic substances into electricity by anaerobic oxidation (Civan et al., 2021; Naik and Jujjavarappu, 2020). Sustainable wastewater treatment methods are being developed by the application of both chemical and biological catalytic treatment methods, and alternative solutions to water scarcity are proposed by wastewater remediation and increasing its reusability. Application of sustainable wastewater treatment methods is of importance as a part of circular economy defined as a sustainable economic model aiming to utilize recourses effectively and to minimize waste, emission, and energy loss by reducing primary sources and avoiding resource depletion (Kurniawan et al., 2021).

5.2 WATER CHARACTERISTICS

The characteristics of water are generally classified under three main groups including physical, chemical, and biological characteristics. The most common water characteristics are presented in Figure 5.1.

5.2.1 PHYSICAL CHARACTERISTICS OF WATER

The physical characteristics of water are easy to measure and some of them could be noticed by the senses of touch, sight, smell, and taste. The physical characteristics of water comprise temperature, taste, odor, color, solids, turbidity, conductivity, and salinity (Liu and Liptak, 1999; Tomar, 1999). The physical, chemical, and biological characteristics of water are listed in Figure 5.1.

5.2.1.1 Temperature

Water temperature influences some of the important physical properties of water including thermal capacity, density, specific weight, viscosity, surface tension, conductivity, and solubility of dissolved gases. Aquatic organisms need specific temperature conditions to live and reproduce. If the water temperature is too high, the organisms cannot survive or are forced to migrate to new locations. As the temperature of water rises, the metabolism of fish and other aquatic organisms increases and the solubility of oxygen decreases, which is affecting the aquatic life negatively. In addition, the increase in water temperature amplifies odors and tastes (Dzurik et al., 2018).

5.2.1.2 Taste and Odor

he taste and odor of water occur due to the dissolved impurities, often organic in nature (phenols, chlorophenols, etc.). Odor can be caused by the decomposition of vegetable matter or microbial activity in wastewaters. It is measured by odor threshold tests in which the water samples are diluted with odor-free water successively until the odor is no longer detectable, whereas taste is evaluated using three methods (De Zuane, 1996; Tebbutt, 1998):

- Flavor threshold test (FIT): Water samples are diluted with increasing amounts of reference water until a panel of testers concludes that there is no perceptible flavor.
- Flavor rating assessment (FRA): Testers are asked to rate the flavor from very favorable to very unfavorable.
- Flavor profile analysis (FPA): Measures both the taste and odor of a water sample in comparison to taste and odor reference standards.

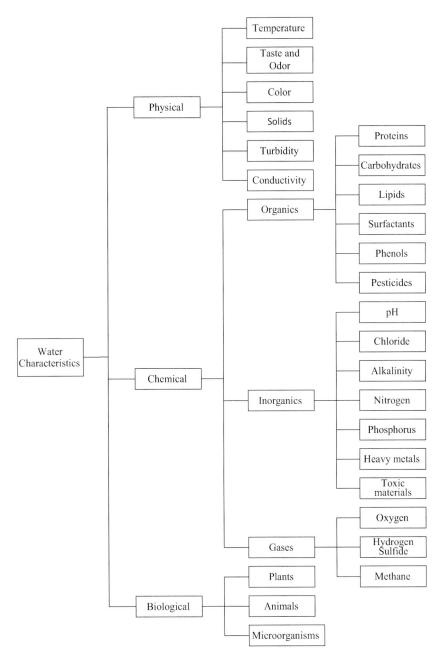

FIGURE 5.1   Physical, chemical, and biological characteristics of water.

The chlorine taste can be eliminated by using a simple activated carbon filter, whereas the sulfide taste and odor can be removed through aeration purging or chlorine oxidation (Cotruvo, 2018).

5.2.1.3 Color

Color in water is important from the viewpoint of aesthetics besides the health effect. Intense color in water is generally caused by the presence of iron, manganese, humus, peat, and weeds. The color of water is classified as true color or apparent color according to the source of the color (De Zuane, 1996).

- Apparent color: Occurs due to the suspended materials
- True color: Occurs due to the dissolved materials and is obtained by filtering or centrifuging the samples.

Color can be measured by comparison with potassium chloroplatinate (Brezonik et al., 2015) standards or by scanning at different spectrophotometric wavelengths.

5.2.1.4 Turbidity

Turbidity refers to the clarity of the water and occurs due to the suspended and colloidal materials. Slit, clays, and other tiny inorganic materials, algae, and organic substances cause turbidity. The small suspended particles scatter or absorb the light. The reduction of light penetration influences the photosynthesis and growth of aquatic life negatively (Dzurik et al., 2018). Turbidity is determined by turbidimeters that measure the intensity of scattered light. Coagulation, filtration, and sedimentation can be applied for turbidity removal (Cotruvo, 2018).

5.2.1.5 Solids

Both natural water and wastewaters contain many dissolved and suspended solids. The solid content of water is generally classified as dissolved, suspended, settleable, and colloidal solids according to their sizes (Spellman, 2013).

- Total dissolved solids (TDS): Soluble materials in water (solids pass through a filter).
- Suspended solids (SS): Solids that can be filtered out.
- Settleable solids: Settle under the influence of gravity (e.g., silt and heavy organic solids).
- Colloidal solids: Extremely fine particles that will not settle for days or weeks, their diameters are <1 μm.

Total suspended solid is significant for designing wastewater treatment systems and the concentration of total dissolved solids is an important factor for various water uses.

5.2.2 Chemical Characteristics of Water

The main chemical characteristics of water comprise pH, alkalinity, acidity, hardness, oxidation–reduction potentials, dissolved oxygen, oxygen demand comprising the biochemical oxygen demand (BOD) and chemical oxygen demand (COD); total organic carbon, nitrogen, and phosphate content; chloride, trace element, and trace organic content.

5.2.2.1 pH

The pH is the measure of hydrogen ion concentration in a solution. An excess amount of hydrogen ions makes the solution acidic, whereas an excess of hydroxide ions makes the solution basic. Many of the chemical reactions are influenced by pH remarkably and the biological activity is usually limited to a narrow pH range of 5–7. In wastewater treatment and corrosion control, pH is important. The pH is commonly monitored by digital pH meters consisting of a potentiometer, a glass electrode, and a reference electrode. Unsuitable pH can cause corrosion problems. In addition, organisms are sensitive to pH changes and only a few aquatic organisms tolerate waters with a pH less than 4 or greater than 10. Therefore, it is important to control the pH of the treated wastewaters prior to discharge into the environment (Weiner and Matthews, 2003).

5.2.2.2 Alkalinity

Alkalinity is the capacity of water to accept H^+ ions. Generally, the basic molecules responsible for alkalinity in water are bicarbonate ions, carbonate ions, and hydroxide ions:

$$HCO_3^- + H^+ \rightarrow CO_2 + H_2O$$

$$CO_3^- + H^+ \rightarrow HCO_3^-$$

$$OH^- + H^+ \rightarrow H_2O$$

Minor contributors to alkalinity are ammonia and the conjugate bases of phosphoric, silicic, boric, and organic acids. At pH values below 7, [H^+] in water detracts significantly from alkalinity.

Alkalinity and the basicity are different concepts, for instance (Manahan, 2010):

- 10^{-3} M NaOH solution: pH = 11, neutralizes 10^{-3} M of acid.
- 0.1 M HCO_3^- : pH = 8.34, neutralizes 0.1 M of acid.

Alkalinity is measured by determining the amount of acid needed to lower the pH in a water sample to a specific endpoint, and the results are generally reported in standardized units as mg $CaCO_3$/L (Weiner and Matthews, 2003).

5.2.2.3 Acidity

Acidity is the capacity of the water to neutralize OH^-. Generally, it results from the presence of weak acids, especially carbon dioxide, but sometimes comprises others such as $H_2PO_4^-$, H_2S, ferric ion, proteins, and fatty acids. Because of the dissolved CO_2, pure water is slightly acidic with a pH less than 7. Many of the natural waters and domestic sewage are buffered by carbon dioxide/bicarbonate systems. Carbon dioxide acidity is in the pH range of 4.5–8.2, whereas mineral acidity caused by industrial pollutants occurs below pH 4.5 (Manahan, 2010; Tebbutt, 1998).

5.2.2.4 Hardness

Hardness of water is defined as the capacity of water sample to precipitate soap. The presence of calcium and magnesium ions and polyvalent ions, such as iron and aluminum, also cause hardness. Hardness is expressed as mg $CaCO_3$/L. There are two types of hardness comprising carbonate and noncarbonated hardness:

- Carbonate hardness: Occurs due to metals associated with HCO_3^-. It is equal to the summation of the carbonate alkalinity and bicarbonate alkalinity.
- Noncarbonate hardness: Occurs due to metals associated with SO_4^{2-}, Cl^-, and NO_3^-. It is evaluated by subtracting carbonate hardness from total hardness.

Hardness is calculated from the concentration of calcium and magnesium ions in the sample or by titrating with EDTA (Patnaik, 2017).

5.2.2.5 Oxidation–Reduction Potentials (ORP)

Oxidation–Reduction Potentials (ORP) is the tendency of a chemical matter to oxidize or reduce another chemical. It is the measured in millivolts by using an ORP sensor consisting of an ORP electrode and a reference electrode. ORP depends on the chloride ion and pH (hydrogen ion) present in the water (Li and Liu, 2019).

5.2.2.6 Dissolved Oxygen (DO)

Oxygen is fundamental to aquatic life though it is poorly soluble in water. Dissolved oxygen concentration is inversely proportional to temperature, and the maximum amount of oxygen dissolved in water at 0°C is 14.6 mg/L. Chlorides in water decrease the solubility of oxygen. In addition, oxygen demand of the organic pollutants removed the oxygen. Dissolved water may be determined by using either electrode or iodometric titration by using $MnSO_4$ and NaOH (Winkler method) (Weiner and Matthews, 2003).

5.2.2.7 Biochemical Oxygen Demand (BOD)

As one of the key water quality parameters, biochemical oxygen demand (BOD) is defined as the amount of oxygen per unit volume of water required by microorganisms to stabilize the oxidizable matter in water. BOD is determined by measuring the oxygen required by bacteria or other microorganisms to decompose the organic substances in a water sample. It is an indirect measure of organic matter amount in the sample and expresses how much dissolved oxygen can be removed from water during the decomposition of organic matter. BOD tests are performed over specific time intervals such as 5, 10, 20, and 30 days (Davie, 2008; Stoddard et al., 2003).

The 5-day BOD (BOD_5) is the most widely used BOD test. A 5-day BOD test (BOD_5) begins by placing water or effluent samples into two standard BOD bottles. One of them is analyzed immediately to measure the initial dissolved oxygen concentration in the effluent. The second BOD bottle is sealed and stored at 20°C in the dark. After 5 days, the amount of dissolved oxygen remaining in the sample is measured. The difference between the initial and ending oxygen concentrations is the BOD_5.

5.2.2.8 Chemical Oxygen Demand (COD)

Chemical oxygen demand (COD) is the oxygen equivalent of organic matter in the water sample that is sensitive to oxidation by a strong oxidizing agent. Generally, the boiling mixture of potassium dichromate $(K_2Cr_2O_7)$–H_2SO_4 is used to determine chemical oxygen demand since it can oxidize most types of organic matter. Other strong oxidants, such as potassium permanganate–sulfuric acid, are also effective. Silver is used as the catalyst.

In order to determine COD, a certain amount of $K_2Cr_2O_7/H_2SO_4$ solution is added to a known amount of sample and the mixture is boiled. The oxidizing agent Cr^{VI} is reduced to Cr^{III}. The amount of consumed Cr^{VI} is evaluated by titration or colorimetry. The difference between the initial and remaining amounts of hexavalent chromium is proportional to the COD (Patnaik, 2017).

5.2.2.9 Total Organic Carbon (TOC)

Determination of the total organic carbon levels is a significant water quality criterion since the identified compounds are approximately 5–10% of the contaminants that can be found in the polluted wastewaters (Greyson, 2020). The terms related to the carbon content of the water are listed as follows:

- Total carbon (TC): All the carbon in the sample.
- Total inorganic carbon (TIC): Carbonate, bicarbonate, and dissolved CO_2.
- Total organic carbon (TOC): Caused by decaying vegetation, bacterial growth, metabolic activities, and chemicals.
- Non-purgeable organic carbon (NPOC): Organic carbon remaining in an acidified sample after purging the sample with gas.
- Purgeable (volatile) organic carbon (VOC): Organic carbon that has been removed from a neutral or acidified sample by purging with an inert gas.
- Dissolved organic carbon (DOC): Organic carbon remaining in a sample after filtering the sample.

- Suspended organic carbon (particulate organic carbon, POC): The carbon in particulate form that is too large to pass through a filter.

Total organic carbon is measured by oxidizing the organic carbon to CO_2 and H_2O and measuring the CO_2 gas using an infrared carbon analyzer. The oxidation occurs by direct injection of the sample into a high-temperature (680–950°C) combustion chamber or by placing a sample into an oxidizing agent such as potassium persulfate, sealing, and heating the sample to complete the oxidation.

5.2.2.10 Nitrogen Content

Nitrogen is necessary for biological reactions and exists in five main forms (Vesilind and DiStefano, 2006):

- Organic nitrogen: Proteins, amino acids, urea
- Ammonia: Ammonium salts, e.g., $(NH_4)_2CO_3$ or free ammonia
- Nitrite (NO_2): Intermediate oxidation stage
- Nitrate (NO_3): Final oxidation product of nitrogen
- Dissolved nitrogen gas

Typically, municipal wastewater contains 60% ammonium nitrogen and 40% organic nitrogen with negligible concentrations of nitrate or nitrite. Untreated domestic wastewater typically contains 20–50 mg/L of total nitrogen (Liu and Liptak, 1997).

Ammonia together with organic nitrogen is considered an indicator of pollution. These two forms of nitrogen are often combined in one measure, called Kjeldahl nitrogen. Aerobic decomposition eventually generates nitrite and finally nitrate. The organic nitrogen can be in the form of humus or intermediate products of degradation of organic substances. The nitrogen is converted into different forms in the nitrogen cycle. The important processes in the nitrogen cycle include fixation, ammonification, nitrification, and denitrification. Nitrogen in water is measured by colorimetric techniques. Ammonia can be determined by adding Nessler reagent into the sample (Barbooti, 2015; Vesilind et al., 2013).

5.2.2.11 Phosphate Content

High phosphorus content, as phosphates, together with nitrates and organic carbon are usually associated with excess growth of aquatic plants. Fertilizers and detergents are major sources of phosphates. Uncontaminated waters contain 10–30 µg/L total phosphorus. Most wastewaters contain sufficient nitrogen and phosphorus to support massive algal blooms (Nemerow et al., 2009).

5.2.2.12 Trace Elements and Trace Organic Content

Trace elements are detected at very low levels of a few parts per million or less in a given system. Some of the trace elements are recognized as nutrients required by animals and plants, including some that are essential at low concentrations but toxic at higher levels. Arsenic, beryllium, boron, chromium, copper, fluoride, iodine, iron, lead, manganese, mercury, molybdenum, selenium, and zinc are among the trace elements present in natural waters (Manahan, 2010). The common sources of these trace elements are mining byproducts, industrial wastes, metal plating, and detergents. The trace organics are released into the water sources by human activities and industrial wastes. The most common hazardous organic compounds found in wastewaters are benzene, chlorophenols, estrogens, pesticides, polynuclear aromatic carbons, and trihalomethanes. There is a concern about the possible health effects in case these chemicals are consumed over a long lime even at trace levels (Tebbutt, 1998).

5.2.3 Biological Characteristics of Water

The biological impurities in water comprising algae, fungi, and pathogenic microorganisms present in water naturally due to the release of improperly treated industrial and domestic wastewaters into the environment influence the biological characteristics of water and adversely affect human health as well as aquatic flora and fauna. Water should be free from all kinds of bacteria, viruses, protozoa, and algae. The coliform and *Escherichia coli* count should be zero (Agarwal, 2019).

Examples of common waterborne pathogens include bacteria (e.g., *Campylobacter, Clostridium botulinum, Clostridium perfringens, E. coli, Legionella, Salmonella paratyphi, Salmonella typhi, Shigella, Staphylococcus aureus, Vibrio comma, Yersinia enterocolitica*), protozoans (e.g., *Cryptosporidium, Entamoeba histolytica, Giardia lamblia*), and viruses (e.g., Hepatitis A, Poliovirus). The microorganisms present in water can cause numerous diseases such as gastroenteritis, pneumonia-like pulmonary, paratyphoid, typhoid fever, shigellosis, cholera, cryptosporidiosis, dysentery, giardiasis, and hepatitis (Weiner and Matthews, 2003).

5.3 WASTEWATER TREATMENT METHODS

5.3.1 Conventional Wastewater Treatment Methods

Conventional wastewater treatment systems consist of a combination of physical (e.g., screening, sedimentation, aeration, flotation, equalization, adsorption), chemical (e.g., chlorination, neutralization, coagulation, precipitation, ion exchange), and biological (e.g., activated sludge, aerobic or anaerobic digestion, trickling filtration) processes to remove solids, organic substances, and nutrients from wastewater. The main target of preliminary treatment is the removal of coarse solids and large materials such as wood, cloth, paper, plastics, garbage, and fecal matter. In the primary treatment, organic and inorganic solids are separated by sedimentation and flotation. Secondary treatment aims to remove organic material through biological processes, whereas tertiary treatment aims to remove nutrients such as nitrogen and phosphorous, remaining suspended solids, toxic and persistent organic compounds that could not be eliminated in the previous stages (Mantzavinos, 2007; Sonune and Ghate, 2004). In the following, the operating principles of some of the commonly used traditional wastewater treatment methods are summarized.

5.3.1.1 Sedimentation

The suspended matter in textile effluent can be removed efficiently and economically by sedimentation. The sedimentation tanks are designed to enable smaller and lighter particles to settle under gravity. Sedimentation is applied for the grit removal, organic matter removal from the sewage inlet (primary sedimentation), the separation of biomass from biological treatment process (secondary sedimentation), and the sludge thickening. The most common equipment used includes rectangular and circular clarifiers. This process is useful for treatment of wastes containing high percentage of settable solids or when the waste is subjected to combined treatment with sewage (Matko et al., 1996; Nemerow, 2007).

5.3.1.2 Equalization and Neutralization

Industrial wastewaters require a pretreatment step to provide a consistent water flow. Flow equalization is applied to minimize the variability of wastewater flow rates and composition. The wastewater influent enters the equalization basin first before it is sent to the rest of the treatment process. In some cases, air is used to mix the slurry in the container to prevent it from sinking to the bottom during the equalization. If a wastewater stream is hazardous due to corrosivity, neutralization is necessary before proceeding to the next treatment stages. In the neutralization process, the pH of water is adjusted by the addition of an acid or a base, depending on the process requirements (Azanaw et al., 2022; Goel et al., 2005).

5.3.1.3 Coagulation/Flocculation

Coagulation and flocculation are widely used to remove a broad range of contaminants from wastewaters, including colloidal particles and dissolved organic materials treat wastewaters. The most commonly used coagulants are aluminum and iron salts. There are many factors such as coagulant/flocculant type and dosage, mixing speed and time, temperature, and retention time influencing the efficiency of coagulation and flocculation (Khouni et al., 2011). The coagulation mechanism can be explained as the interaction of coagulants with colloids through either charge neutralization or adsorption to large flocs, which can be separated by sedimentation or flotation. Aluminum salts coupled with polymeric flocculants have been extensively used for the treatment of industrial wastewaters (Ansari et al., 2018).

5.3.1.4 Ion Exchange

In an ion-exchange process, a reversible ion interchange between the liquid and solid phases occurs, where an insoluble resin separates ions from an electrolytic solution and delivers other ions of like charge in a chemically equivalent amount without any structural variation of the resins. Ion exchange can also be used for the recovery of valuable metal ions from inorganic wastewaters (Kurniawan et al., 2006). Cation resins have fixed negatively charged groups and just allow cations to pass through and reject anions. They release positive ions such as H^+ or Na^+ in exchange for impurity cations. In contrast, anion resins have fixed positively charged groups that accept only anions to pass through and reject cations. They release negative ions such as OH^- or Cl^- in ion exchange for impurity anions (Din et al., 2021).

5.3.1.5 Adsorption

Adsorption is extensively used for the removal of a wide variety of pollutants from wastewaters. It is a mass transfer process in which a solute is transferred from liquid phase to the surface of a solid (adsorbent) and bounded by physical or chemical interactions. When the wastewaters contact with an adsorbent with porous surface structure, liquid–solid intermolecular forces of attraction result in deposition of the contaminants at the surface of the solid. Larger surface areas of adsorbents lead to higher adsorption capacity and surface reactivity. Different types of low-cost materials as adsorbents such as tree bark, wood charcoal, saw dust, alum sludge, red mud, peanut hulls, peat, corncobs, cocoa shells, and other waste materials can be used for the adsorption of some toxic substances (Chiban et al., 2012; Rashed Nageeb, 2013).

5.3.1.6 Activated Sludge Process

The activated sludge process is one of the most commonly applied biological wastewater treatment methods. In this process, a suspension of bacterial biomass is responsible for the removal of biological nitrogen and phosphorus and organic carbon. Activated sludge consists of many microorganisms, immobilized in extracellular polymeric materials or polymeric matrices constituting proteins, polysaccharides, humic acids, and lipids. The composition of the wastewater, pH, temperature, and the presence of toxic compounds affects the performance of activated sludge process remarkably (Gernaey et al., 2004; Ni and Yu, 2012).

5.3.2 ADVANCED OXIDATION PROCESSES

Conventional wastewater treatment methods are mostly ineffective for the elimination of refractory compounds present in wastewaters. Physical treatment methods transfer the pollution from one phase to another instead of mineralization of the contaminants. The solid wastes generated in conventional wastewater treatment systems require further treatment, thus increasing the total cost. Sludge generation and regeneration of adsorbents or resins are the limitations of the conventional wastewater treatment methods. The bottleneck of the conventional methods can be overcome by the application of environmentally friendly advanced oxidation processes (AOPs).

Advanced oxidation technologies have proven to be successful in water purification. In these processes, emerging pollutants and toxic and harmful contaminants are decomposed effectively. AOPs comprise a variety of methods for producing hydroxyl radicals ($^{\bullet}OH$), reactive oxygen species such as superoxide anion radical $\left(O_2^-\right)$, and singlet oxygen. Hydroxyl radicals are the strongest oxidants that can react with most of the organic compounds present in the wastewaters, often at a diffusion-controlled reaction rate. They react unselectively once formed and pollutants will be quickly fragmented and converted into small inorganic compounds. The reactive species are generated from primary oxidants (e.g., ozone, peroxymonosulfate, hydrogen peroxide, oxygen), energy sources (e.g., ultraviolet or visible light), and various catalysts. Optimized dosages, sequences, and combinations of the oxidants and catalysts lead to a maximum •OH yield (Moradi et al., 2018; O'Shea and Dionysiou, 2012). The major categories of advanced oxidation processes are chemical, photochemical, sonochemical, and electrochemical processes. In addition, miscellaneous methods including non-thermal plasma, supercritical water oxidation, and microwave-assisted processes are categorized under advanced oxidation processes (Gautam et al., 2019; Valange and Védrine, 2018). The classification of the advanced oxidation processes is shown in Figure 5.2.

The main advantages of the advanced oxidation technologies are (Coha et al., 2021) as follows:

- Applicable to a wide variety of wastewaters with different compositions due to nonselective oxidation reactions
- Removal of both organic and inorganic compounds (e.g., cyanides, sulfides, sulfites, nitrites)
- Decomposition of compounds responsible for odor, taste, and color effectively
- Ability to be operated under ambient conditions (atmospheric pressure and room temperature)
- High quality of the treated water due to the mineralization of pollutants
- Increasing the biodegradability of the contaminants present in raw wastewater
- Enabling the reuse of wastewaters

In the following, some of the most commonly used advanced oxidation processes applied for wastewater reclamation and reuse are introduced.

5.3.2.1 Fenton and Fenton-Like Oxidation Process

Among various advanced oxidation methods, Fenton-like oxidation is one of the most efficient and low-cost methods to remove recalcitrant organic compounds. The oxidation–reduction reactions between ferric and ferrous ions take place in the presence of H_2O_2, which promote the generation of reactive species such as hydroxyl (•OH) and hydroperoxyl (HO$_2$•) radicals. The hydroxyl radicals with high oxidation potential have the ability of degrading pollutants non-selectively. Fenton oxidation is reported as a promising method for the treatment of industrial wastewaters containing non-biodegradable organic substances. In the following, the degradation mechanism of organic contaminants by Fenton oxidation is shown (Nidheesh et al., 2013):

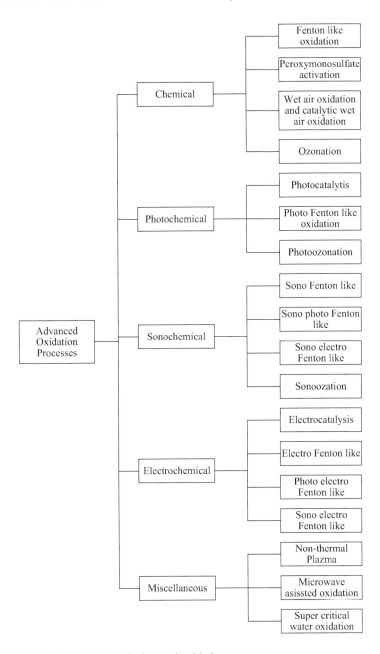

FIGURE 5.2   Classification of advanced oxidation processes.

$$Fe^{2+} + H_2O_2 \rightarrow Fe^{3+} + OH^- + {}^\bullet OH$$

$$RH + {}^\bullet OH \rightarrow R^\bullet + H_2O$$

$$R^\bullet + Fe^{3+} \rightarrow R^+ + Fe^{2+}$$

$$Fe^{2+} + {}^\bullet OH \rightarrow Fe^{3+} + OH^-$$

Fe^{3+} ions are regenerated in the Fenton process by the reaction between Fe^{2+} and hydrogen peroxide:

$$Fe^{3+} + H_2O_2 \rightarrow Fe^{2+} + H^+ + HO_2{}^\bullet$$

$$Fe^{3+} + HO_2{}^\bullet \rightarrow Fe^{2+} + H^+ + O_2$$

Despite its high efficiency, the homogeneous Fenton-like oxidation process has important following disadvantages, which can be overcome using heterogeneous Fenton-like oxidation (Hassan and Hameed, 2011; Liu et al., 2018; Singh et al., 2016):

- Iron ions must be separated from the reaction medium by precipitation, which is expensive in labor, reagents, and time.
- It is limited by a narrow pH range (pH 2–3).
- Iron ions are deactivated due to complexion with several reagents such as phosphate anions and intermediate oxidation products.

5.3.2.2 Photo-Fenton Oxidation and Photocatalytic Degradation Processes

Photo-Fenton oxidation is a combination of Fenton reagents (H_2O_2 and Fe^{2+}) and UV–visible irradiation that improves the formation of hydroxyl radicals via photoreduction of Fe^{3+} to Fe^{2+} ions and peroxide photolysis at shorter wavelengths.

$$Fe^{3+} + H_2O + UV \rightarrow OH + Fe^{2+} + H^+$$

$$Fe\,(OH)_2{}^+ + h\nu \rightarrow Fe^{2+} + {}^\bullet OH$$

$$H_2O_2 + h\nu \rightarrow 2{}^\bullet OH$$

The photogenerated ferrous ions enter the Fenton reaction to produce supplemental hydroxyl radicals. Since the concentration of ferrous ions is increased, the overall reaction is accelerated. Consequently, the oxidation rate of photo-Fenton oxidation is higher than the rate of Fenton oxidation. In addition, there is a considerable decrease in iron requirement and in sludge generation in photo-Fenton oxidation. Furthermore, using solar or UV light has important influences on inactivation of microorganisms in contaminated water to increase the reusability of the treated water for irrigation or drinking water purposes (Malik and Saha, 2003; Rahim Pouran et al., 2015).

Among various advanced oxidation technologies, photocatalytic oxidation has received great attention as one of the most viable environmental remediation methods; yet the cost competitiveness of photocatalysis is still lower than traditional wastewater treatment processes and some advanced oxidation processes (Park et al., 2013).

An efficient light harvesting and a redox process based on the photogenerated charge carriers are essential factors for an efficient photocatalytic oxidation process. Semiconductor photocatalysts with appropriate band gaps are accepted as light-harvest medium with photogenerated charge carriers (Sang et al., 2015).

When a photocatalyst absorbs light, electrons are promoted from the valence band to the conduction band and electron–hole pairs are generated.

$$\text{Photocatalyst} + h\nu \rightarrow e_{cb}^- + h_{vb}^+$$

Both of these entities migrate to the surface of the photocatalyst to enter into a redox reaction with other species on the catalyst surface. Generally, holes react with surface-bound water to generate hydroxyl radicals while electrons react with oxygen to generate superoxide radical anions.

$$H_2O + h_{vb}^+ \rightarrow \ OH + H^+$$

$$O_2 \rightarrow e_{cb}^- + O_2^-$$

These reactions prevent the recombination of the electron–hole pairs. Produced radicals are responsible for the removal of water pollutants since they react with the organic compounds to degrade them (Rauf et al., 2011).

5.3.2.3 Wet Air Oxidation (WAO) and Catalytic Wet Air Oxidation (CWAO) Processes

Wet oxidation is a hydrothermal process in which high-pressure air or oxygen as oxidant is used at high temperatures. Wet air oxidation is a suitable option for treating industrial wastewaters that are too dilute to incinerate and too concentrated for biological treatment. The suspended or dissolved substances in a liquid phase are oxidized with dissolved oxygen. The wet air oxidation chemistry involves chain reactions of radicals generated from organic and inorganic compounds present in the reaction medium. The organic contaminants are converted into biodegradable substances or mineralized completely, whereas the inorganic compounds are converted into their form with higher oxidation value (e.g., conversion of sulfides into sulfates) (Tungler et al., 2015).

Wet air oxidation process breaks down refractory pollutants into simpler compounds prior to their release into the environment. Generally, this aqueous-phase flameless combustion occurs under high temperatures (473–593 K) and pressures (20–200 bar) by means of active oxygen species, such as hydroxyl radicals. Residence times varied between 15 min and 120 min, and the typical extent of chemical oxygen demand removal changes between 75% and 90% (Ovejero et al., 2011).

The degradation mechanism of wet oxidation is complex; however, generally the chemical reaction stage occurs mostly via free-radical reactions including initiation, propagation, and termination reactions. The main reactions proposed for the wet air oxidation of organic compounds (RH) are listed as (Zhou and He, 2007):

$$RH + O_2 \rightarrow HO_2^{\bullet} + R^{\bullet}$$

$$H_2O + O_2 \rightarrow HO_2^{\bullet} + {}^{\bullet}OH$$

$$R^{\bullet} + O_2 \rightarrow ROO^{\bullet}$$

$$ROO^{\bullet} + RH \rightarrow ROOH + R^{\bullet}$$

$${}^{\bullet}OH + RH \rightarrow R^{\bullet} + H_2O$$

$$HO_2^{\bullet} + {}^{\bullet}OH \rightarrow H_2O + O_2$$

Catalytic wet air oxidation is one of the most promising technologies for the treatment of hazardous, toxic, and non-biodegradable wastewaters. Due to the presence of a catalyst, high oxidation rates are achieved under less severe reaction conditions. In catalytic wet air oxidation process, organic compounds are oxidized into harmless products at much lower temperatures and pressures than in uncatalyzed thermal processes. Depending on the type and concentration of organic compounds dissolved in wastewater, this process can be operated either to reduce their concentration or to eliminate them (Levec and Pintar, 2007; Roy and Saroha, 2014). Although homogeneous catalysts such as dissolved metal salts are effective in degradation of organic compounds, an additional separation step is required to remove or recover the metal ions from the treated effluents leading to increase in operational costs. Therefore, development of active heterogeneous catalysts is significant. Various solid catalysts including noble metals, metal oxides, and mixed oxides have been widely used for the catalytic wet air oxidation of pollutants (Kim and Ihm, 2011).

5.3.2.4 Supercritical Water Oxidation

Supercritical water oxidation occurs at a temperature and pressure above the critical point of water (647 K and 22.1 MPa). The physical properties of water change dramatically around the critical point. The unique properties of supercritical water such as low viscosity, high diffusivity, and low dielectric constant provide an appropriate environment for oxidation and gasification (Zhao and Jin, 2020). Many of organic compounds, such as alkanes and aromatics, and gases are completely dissolved in supercritical water, forming a single phase. Reactions in supercritical water proceed by free-radical or ionic mechanisms depending on pressure and temperature. Free-radical reactions dominate at high temperatures while ionic reactions occur at subcritical temperature and very high pressures (Güngören Madenoğlu et al., 2016).

Supercritical water oxidation has shown potential for treating effluents of various industries including pulp and paper, pharmaceutical, textile, pesticides, dairy, petrochemical, explosives, and distillery. Supercritical water oxidation is an attractive alternative to the conventional wastewater treatment methods for the elimination of toxic, persistent, refractory, and hazardous organic pollutants at high concentrations due to the unique reaction features of supercritical water. Though refractory organic compounds can be rapidly oxidized to water, carbon dioxide, and nitrogen, technical challenges including corrosion, salt deposition, clogging, and high operating costs limit the application of supercritical water oxidation (de Souza et al., 2022; Gong et al., 2008).

5.3.2.5 Electrochemical Advanced Oxidation Processes (EAOPs)

The electrochemical oxidation methods have been found to be a promising technology for environmental remediation and removal of organic pollutants. Electrochemical advanced oxidation methods are based on the in situ generation of hydroxyl radicals, which reacts with a wide variety of organic pollutants up to their mineralization to carbon dioxide, water, and inorganic ions due to its high standard redox potential. The most powerful electrochemical advanced oxidation processes generate hydrogen peroxide from the two-electron reduction of injected O_2:

$$O_2 + 2H^+ + 2e^- \rightarrow H_2O_2$$

The interest of using electrochemical advanced oxidation processes is based on their ability of reacting on pollutants by using both direct and indirect influences of electrical current. Direct oxidation can be accomplished through mineralization with hydroxyl radicals generated at the electrode surface (Daghrir et al., 2014; Pipi et al., 2014).

5.3.2.6 Sonocatalysis

Sonocatalysis is one of the most beneficial advanced oxidation processes for the remediation of wastewaters. The effect of ultrasound on heterogeneous catalytic reactions is attributed to both in situ generation of chemically active species and mechanical influence of acoustic cavitation resulting in dispersion of powdered catalyst and removal of the passivating layer from catalyst surface (El Hakim et al., 2021). Nucleation, cavitation, bubble dynamics/interactions, thermodynamics, and chemical processes are the mechanisms of sonolysis. Ultrasonic irradiation decreases mass transfer limitations by increasing the turbulence in water. It enhances catalytically active surface area via the deagglomeration of the particles (Pradhan et al., 2019).

Ultrasonic irradiation process is proven to be capable of degradation of various recalcitrant organic pollutants comprising phenolic compounds, chloroaromatic compounds, polycyclic aromatic hydrocarbons, pesticides, and dyes. Safety, cleanliness, high removal efficiencies, and low energy requirement are the main advantages of using sonocatalytic oxidation. In addition, sonocatalysis can be applied without any generation of secondary pollutants (Pang et al., 2011). Since ultrasound alone is insufficient for complete mineralization of pollutants in water, various catalysts have been introduced to reduce energy consumption and improve the efficiency of ultrasonic processes (Chong et al., 2017).

5.3.2.7 Ozonation

Ozone either reacts directly with organic pollutants or decomposes into highly reactive species. In basic medium, ozone decomposes quickly to yield hydroxyl radicals and other reactive species:

$$O_3 + OH^- \rightarrow O_3^{\bullet -} + {}^\bullet OH$$

$$O_3^{\bullet -} \rightarrow O^{\bullet -} + O_2$$

$$O^{\bullet -} + H+ \rightarrow {}^\bullet OH$$

In acidic medium, ozone is stable and may react with organic substrates directly. Ozonation has advantages for the removal of organic compounds in wastewaters (Wu et al., 2008):

- No sludge formation
- Safe and easy operation
- Degradation occurring in one step
- Requirement of little space
- Decomposition of the residual ozone into oxygen and water

The performance of the ozonation process can be improved by the addition of H_2O_2 or peroxymonosulfate and using UV light irradiation since they will decompose ozone rapidly to produce hydroxyl or sulfate radicals ($SO_4^{\bullet -}$) (Song et al., 2022).

The recent studies on application of various advanced oxidation methods for the remediation of synthetic or real wastewaters are summarized in Table 5.1.

5.3.3 MICROBIAL FUEL CELL

Microbial fuel cells are systems in which electricity is produced by transferring electrons from protons obtained by anaerobic oxidation of various substrates such as glucose, acetate, starch, and cellulose in the anode compartment to the cathode compartment with the help of a resistor (Zhu and Ni, 2009). The protons (hydrogen ions) released in the anode compartment, as well as the electrons, reach the cathode compartment by diffusion through a proton-exchange membrane separating the anode and cathode compartments. Hydrogen transferred to the cathode chamber is converted to water by combining with a final electron acceptor, where oxygen is generally preferred. Anode compartment where anaerobic bacteria are located should be free of oxygen or other electron acceptors. In case of using

TABLE 5.1

Recent Studies on Use of Advanced Oxidation Methods for the Treatment of Industrial Wastewaters

Treatment Method	Wastewater Type/Target Pollutant	Catalyst	Reaction Conditions	Pollutant Removal	Reference
CWAO	Reactive Red 2	Copper oxide loaded activated carbon (CuO/AC)	6 g/L of catalyst loading, 120°C, 3 bar of oxygen partial pressure, 120 min reaction time, 400 rpm, C_0=100 mg/L	100% decolorization and 88.57% COD removal	Ayalkie Gizaw and Gabbiye Habtu (2022)
Fenton-like oxidation	Dyeing wastewater	Pyrite	10 mM H_2O_2 and 24 g/L catalyst loading, pH 3–6, 180 min of reaction time	76.67% COD removal and 97.09% decolorization	Chen et al. (2021)
CWAO	Phenol	Electron-enriched Pt/ TiO_2	100°C and 2 MPa O_2, 400 rpm stirring rate, C_0=1,000 mg/L, 4 g/L catalyst loading	88.8% TOC removal	Fu et al. (2022)
Fenton-like oxidation	Landfill leachate	Cu/ZrO_2	70°C, pH 5, 200 mg/L of catalyst, H_2O_2 30 mL/L, 150 min reaction time	92% TDOC removal	Hussain et al. (2022)
Catalytic ozonation	Methylene blue	Iron-coated reduced graphene oxide (Fe-RGO)	pH 7, catalyst dose of 0.02 g, ozone dose 0.5 mg/min, 10 min of reaction time	96% removal	Qazi et al. (2022)
CWAO	N-methyldiethanolamine (MDEA)	Activated carbon	C_0=18.92 g/L, 230°C, 5.0 MPa, 90 min residence time	98% MDEA removal and ~90% COD removal	Zhao et al. (2022)
Photocatalytic oxidation, CWAO	Pharmaceutical wastewater	Fe/activated carbon	Photocatalytic oxidation: pH 4.5, 2.0 g/L, $[H_2O_2]_0$=0.32 mM. CWAO: pH 3, 50°C, 3 g/L catalyst loading	Photocatalytic oxidation: 72.7% COD removal, CWAO: 83.1% COD removal	Berkin Olgun et al. (2021)

(Continued)

TABLE 5.1 (CONTINUED)
Recent Studies on Use of Advanced Oxidation Methods for the Treatment of Industrial Wastewaters

Treatment Method	Wastewater Type/Target Pollutant	Catalyst	Reaction Conditions	Pollutant Removal	Reference
CWAO	Ammonia	Ru@TiO_2	2 MPa at 190°C in 6 h, C_0=200 mg/L, 250 rpm stirring rate	95.5% conversion	Gai et al. (2021)
CWAO	N-(phosphonomethyl)glycine (PMG)	Fe-CNF/ACB	C_0=100 mg/L, 220°C 25 bar, 6 h, 0.75 g/L catalyst loading	80% COD removal	(Gupta and Verma, 2021)
Photocatalytic oxidation	Methylene blue	TiO_2/graphene oxide	0.2 g/L of catalyst dose, pH 10, $[Cl]_0$= 5 mg/L, 4 h of reaction time	Almost complete dye removal (99%)	Kurniawan et al. (2020)
Photocatalytic oxidation	Ciprofloxacin	TiO_2 nanoparticles immobilized on a glass plate	pH of 5, contact time of 105 min, $[Cl]_0$= 3 mg/L	92.81–86.57% ciprofloxacin removal	Malakootian et al. (2020)
CWAO	Biomethanated distillery wastewater	$FeSO_4 \cdot 7H_2O$	$(COD)_0$ = 40,000 mg/L, 1,200 rpm stirring rate, 0.69 MPa O_2 pressure, 1 h, 175°C, 33 mg/L catalyst loading	~73% COD removal	Bhoite and Vaidya (2018))
CWAO	Paper industry wastewater	CuO-CeO_2	$(COD)_0$=865 ± 32.14 mg/L, $(TOC)_0$=172.3 ± 4.8 mg/L, 1 g/L catalyst loading, 90°C, 2 h	67% COD, 81% color, 64% TOC removal	(Anushree et al. (2017))

Abbreviations: ACB, activated carbon beads; CNF, carbon nanofibers; COD, chemical oxygen demand; CWAO, catalytic wet air oxidation; TDOC, total dissolved organic carbon; TOC, total organic carbon.

anaerobic bacteria in microbial fuel cells, the anode compartment should be operated in completely anaerobic conditions and the cathode compartment in aerobic conditions (Uçar and Toprak, 2014). In addition to energy production in microbial fuel cells, industrial wastes can also be treated and electrical energy can be produced instead of spending energy in the treatment process (Kılıç et al., 2011).

It is possible to apply microbial fuel cells to organic pollutants in hybrid systems formed by Fenton oxidation or photocatalytic oxidation. In a typical electro-Fenton process, H_2O_2 is released by reducing oxygen with electrical energy.

$$O_2 + 2H^+ + 2e^- \rightarrow H_2O_2$$

In microbial fuel cells, on the other hand, H_2O_2 is released by the reduction of oxygen in the cathode section, where the target pollutant is located, with the help of electrons transferred from the anode compartment, without the need for external energy to be supplied to the system (Fu et al., 2010). With the use of iron or iron-containing cathodes in the cathode compartment, the target pollutant can be removed by bio-electro-Fenton oxidation even in neutral conditions without energizing the system (Feng et al., 2010). While iron salts or scrap iron is added to the cathode compartment in homogeneous bio-electro-Fenton systems, carbon nanotube/γ-FeOOH, Fe@Fe$_2$O$_3$/carbon felt, pyrrhotite-coated graphite, etc., composite cathodes are used in heterogeneous systems to reduce iron leaching (Li et al., 2010). In the literature, there are photocatalytic and bio-electro hybrid purification systems in which the cathode reactivity is increased by using rutile TiO$_2$-coated cathodes under visible light for dye removal in microbial fuel cells (Ding et al., 2010).

Thus, microbial fuel cells, in which bacteria are used as biocatalysts, stand out among the alternative methods applied in wastewater treatment with their innovative biotechnological features that enable the production of clean energy directly from biodegradable compounds, decomposition of pollutants, and bioelectricity production.

5.4 REUSE OF WASTEWATERS

The amount of water used increases with the increase in population and the development of industry. Water scarcity is one of the biggest problems of the recent years, and it is estimated that in the coming years, water resources will be one of the most important environmental issues due to the climate change. The scarcity of water resources in the world brings about the improvement and reuse of wastewater. Increased water pressure in terms of both water shortage and water quality causes recycled water as a new water source (Demir et al., 2017).

A significant portion of the water consumption accounts for agricultural irrigation. Regarding the elevating need for crop production and water scarcity, reuse of treated wastewaters for irrigation offers an opportunity in minimizing water stress on freshwater sources by decreasing the quantity of water extracted for irrigation. Reuse of treated wastewaters gains importance particularly in regions of water scarcity (Kesari et al., 2021).

5.4.1 REUSE OF WASTEWATERS IN IRRIGATION

The increasing scarce of freshwater resources and decrease in water quality due to the population growth and industrialization have been becoming an important challenge for agricultural irrigation. Wastewater reuse for irrigation has gained attention recently because appropriately treated wastewaters are ideal for agricultural applications. The quality of the treated wastewaters should be strictly controlled to prevent the soil damage, microbial infection, and accumulation of hazardous chemicals in crops (Jeong et al., 2016).

Due to the difference in water quality requirements of various beneficial uses, the water quality parameters and standards are different. The beneficial use of water includes drinking water, recreational use, fisheries and shellfish production, irrigation, and industrial process water (Gürel and Pehlivanoglu-Mantas, 2010). The assessment of the suitability of the irrigation water is performed based on limiting values of certain physical, chemical, and biological parameters in order to avoid adverse health and environmental effects. The parameters to assess irrigation water quality can be listed as (Phocaides and FAO, 2007):

- Color
- Odor
- Turbidity, NTU
- Biochemical oxygen demand (BOD5), mg/L
- Chemical oxygen demand (COD), mg/L
- Suspended solids (SS), mg/L
- Total coliforms/100 mg
- Fecal coliforms
- Total salinity in terms of electrical conductivity (dS/m) or total dissolved salts (mg/L)
- Acidity/basicity/pH
- Hardness $CaCO_3$ (mg/L)
- Types and concentrations of anions and cations (meq/L)
- Sodium adsorption ratio (SAR)
- Nitrate-nitrogen NO_3-N (mg/L)
- Phosphate-phosphorous, PO_4-P (mg/L)
- Trace elements
- Heavy metals

In literature, application of various wastewater treatment methods for the wastewater reclamation and reuse in irrigation is reported. For instance, Aliste et al. (2022) investigate the reuse of reclaimed agricultural wastewater contaminated with insecticides including chlorantraniliprole, imidacloprid, pirimicarb, and thiamethoxam for growing lettuce in field conditions. Solar photocatalysis was applied in the presence of the $TiO_2/Na_2S_2O_8$ catalyst on a pilot plant to degrade the insecticides their reaction intermediates. Then, reclaimed agro-wastewater was utilized to irrigate lettuce crops grown under greenhouse and agricultural field conditions (Aliste et al., 2022). Aydin et al. (2022) tested the performance of photoelectrocatalytic membrane reactor for the desalination and disinfection of biologically treated textile wastewater with the purpose of reusing it in hydroponic farming. Prior to the use of photoelectrocatalytic membrane reactor, the anionic ion-exchange process was used to eliminate the toxic dyestuff present in the wastewater (Aydin et al., 2022). Pérez-Lucas et al. (2022) studied on detoxification of aqueous solutions of emerging pollutants (pharmaceuticals and pesticides) by using a coupled biological–photocatalytic treatment system. Contaminated water was subjected to biological treatment followed by a photocatalytic process under sunlight for use in crop irrigation (Pérez-Lucas et al., 2022). Bayrakdar et al. (2021) used a sequential photo-Fenton-like oxidation and adsorption process to reusability of treated textile wastewater in irrigation. The agricultural wastes including walnut shell and rice husk were used to prepare activated carbon adsorbent and catalyst supports (Bayrakdar et al., 2021). Bulca et al. (2021) compared the performances of electrocoagulation/adsorption and electrocoagulation/catalytic wet air oxidation hybrid processes to increase the reusability of the treated real textile wastewater for irrigation purpose. Perovskite/rice husk-based activated carbon composite catalyst was used in catalytic wet air oxidation process (Bulca et al., 2021).

By the application of the individual and sequential treatment methods including biological, chemical, and electrochemical processes, the quality of the treated wastewaters could be increased in terms of total organic carbon content, color, turbidity, and suspended solids. Reuse of treated wastewater for irrigation provides a good alternative to reduce the demand for the fresh water resources in agriculture.

5.5 CONCLUDING REMARKS

This chapter reports brief information on application of "Catalytic Methods on Wastewater Remediation" in the light of circular economy.

The authors discuss the water characteristics, fundamentals of wastewater treatment methods, and reuse of wastewaters.

It can be concluded that:

- Application of sustainable wastewater treatment methods is of importance as a part of circular economy.
- Treatment of wastewater and the reuse of these treated streams are very important steps to reach a total circular economy due to the sustainable management of the water and the recoverable/convertible materials in the wastewater.
- Wastewater treatment processes taking into circular economy
  - improves the sustainability of the treatment systems;

- presents benefits from something that has been considered as waste;
- reduces energy consumption;
- reduces the pressure on natural resources regarding water demand.

REFERENCES

Agarwal, S., 2019. *Engineering Chemistry: Fundamentals and Applications*, 2nd ed. Cambridge University Press.

Ahmed, J., Thakur, A., Goyal, A., 2021. *Industrial Wastewater and Its Toxic Effects*, pp. 1–14. https://doi.org/10.1039/9781839165399-00001

Aliste, M., Garrido, I., Hernández, V., Flores, P., Hellín, P., Navarro, S., Fenoll, J., 2022. Assessment of reclaimed agro-wastewater polluted with insecticide residues for irrigation of growing lettuce (*Lactuca sativa L*) using solar photocatalytic technology. *Environ. Pollut.* 292(A). https://doi.org/10.1016/j.envpol.2021.118367

Ansari, S., Alavi, J., Yaseen, Z.M., 2018. Performance of full-scale coagulation-flocculation/DAF as a pre-treatment technology for biodegradability enhancement of high strength wastepaper-recycling wastewater. *Environ. Sci. Pollut. Res. Int.* 25(34), 33978–33991. https://doi.org/10.1007/s11356-018-3340-0

Anushree, Kumar, S., Sharma, C., 2017. Synthesis, characterization and application of $CuO-CeO_2$ nanocatalysts in wet air oxidation of industrial wastewater. *J. Environ. Chem. Eng.* 5(4), 3914–3921. https://doi.org/10.1016/j.jece.2017.07.061

Ayalkie Gizaw, B., Gabbiye Habtu, N., 2022. Catalytic wet air oxidation of azo dye (reactive red 2) over copper oxide loaded activated carbon catalyst. *J. Water Process Eng.* 48, 102797. https://doi.org/10.1016/j.jwpe.2022.102797

Aydin, M.I., Ozaktac, D., Yuzer, B., Doğu, M., Inan, H., Okten, H.E., Coskun, S., Selcuk, H., 2022. Desalination and detoxification of textile wastewater by novel photocatalytic electrolysis membrane reactor for ecosafe hydroponic farming. *Membranes (Basel)* 12(1). https://doi.org/10.3390/membranes12010010

Azanaw, A., Birlie, B., Teshome, B., Jemberie, M., 2022. Textile effluent treatment methods and eco-friendly resolution of textile wastewater. *Case Stud. Chem. Environ. Eng.* 6, 100230. https://doi.org/10.1016/j.cscee.2022.100230

Barbooti, M., 2015. *Environmental Applications of Instrumental Chemical Analysis*. Apple Academic Press. https://doi.org/10.1201/b18376

Bayrakdar, M., Atalay, S., Ersöz, G., 2021. Efficient treatment for textile wastewater through sequential photo Fenton-like oxidation and adsorption processes for reuse in irrigation. *Ceram. Int.* 47(7), 9679–9690. https://doi.org/10.1016/j.ceramint.2020 .12.107

Berkün Olgun, Ö., Palas, B., Atalay, S., Ersöz, G., 2021. Photocatalytic oxidation and catalytic wet air oxidation of real pharmaceutical wastewater in the presence of Fe and $LaFeO_3$ doped activated carbon catalysts. *Chem. Eng. Res. Des.* 171, 421–432. https://doi.org/10.1016/j.cherd.2021.05.017

Bhoite, G.M., Vaidya, P.D., 2018. Improved biogas generation from biomethanated distillery wastewater by pretreatment with catalytic wet air oxidation. *Ind. Eng. Chem. Res.* 57(7), 2698–2704. https://doi.org/10.1021/acs.iecr.7b04281

Brezonik, P.L., Olmanson, L.G., Finlay, J.C., Bauer, M.E., 2015. Factors affecting the measurement of CDOM by remote sensing of optically complex inland waters. *Remote Sens. Environ.* 157, 199–215. https://doi.org/10.1016/j.rse.2014.04.033

Bulca, Ö., Palas, B., Atalay, S., Ersöz, G., 2021. Performance investigation of the hybrid methods of adsorption or catalytic wet air oxidation subsequent to electrocoagulation in treatment of real textile wastewater and kinetic modelling. *J. Water Process Eng.* 40. https://doi.org/10.1016/j.jwpe.2020.101821

Chen, Q., Yao, Y., Zhao, Z., Zhou, J., Chen, Z., 2021. Long term catalytic activity of pyrite in heterogeneous Fenton-like oxidation for the tertiary treatment of dyeing wastewater. *J. Environ. Chem. Eng.* 9(4), 105730. https://doi.org/10.1016/j.jece.2021.105730

Chiban, M., Soudani, A., Sinan, F., Persin, M., 2012. Wastewater treatment by batch adsorption method onto micro-particles of dried *Withania frutescens* plant as a new adsorbent. *J. Environ. Manag.* 95(Supplement), S61–S65. https://doi.org/10.1016/j.jenvman.2011.06.044

Chong, S., Zhang, G., Wei, Z., Zhang, N., Huang, T., Liu, Y., 2017. Sonocatalytic degradation of diclofenac with FeCeOx particles in water. *Ultrason. Sonochem.* 34, 418–425. https://doi.org/10.1016/j.ultsonch.2016.06.023

Civan, G., Palas, B., Ersöz, G., Atalay, S., Bavasso, I., Di Palma, L., 2021. Experimental assessment of a hybrid process including adsorption/photo Fenton oxidation and microbial fuel cell for the removal of dicarboxylic acids from aqueous solution. *J. Photochem. Photobiol. A Chem.* 407, 1–8. https://doi.org/10.1016/j.jphotochem.2020.113056

Coha, M., Farinelli, G., Tiraferri, A., Minella, M., Vione, D., 2021. Advanced oxidation processes in the removal of organic substances from produced water: Potential, configurations, and research needs. *Chem. Eng. J.* 414, 128668. https://doi.org/10.1016/j.cej.2021.128668

Cotruvo, J., 2018. *Drinking Water Quality and Contaminants Guidebook*. CRC Press, Boca Raton, Taylor & Francis. https://doi.org/10.1201/9781351110471

Daghrir, R., Drogui, P., Tshibangu, J., 2014. Efficient treatment of domestic wastewater by electrochemical oxidation process using bored doped diamond anode. *Sep. Purif. Technol.* 131, 79–83. https://doi.org/10.1016/j.seppur.2014.04.048

Davie, T., 2008. *Fundamentals of Hydrology*, 2nd ed. Taylor & Francis.

de Souza, G.B.M., Pereira, M.B., Mourão, L.C., dos Santos, M.P., de Oliveira, J.A., Garde, I.A.A., Alonso, C.G., Jegatheesan, V., Cardozo-Filho, L., 2022. Supercritical water technology: An emerging treatment process for contaminated wastewaters and sludge. *Rev. Environ. Sci. Bio Technol.* 21(1), 75–104. https://doi.org/10.1007/s11157-021-09601-0

De Zuane, J., 1996. *Handbook of Drinking Water Quality*. Wiley. https://doi.org/10.1002/9780470172971

Demir, Ö., Yıldız, M., Sercan, Ü., Arzum, C.Ş., 2017. Atıksuların Geri Kazanılması ve Yeniden Kullanılması. *Harran Üniversitesi Mühendislik Dergisi, Çevre Mühendisliği Bölümü, Şanlıurfa* 2, 1–14.

Din, N.A.S., Lim, S.J., Maskat, M.Y., Mutalib, S.A., Zaini, N.A.M., 2021. Lactic acid separation and recovery from fermentation broth by ion-exchange resin: A review. *Bioresour. Bioprocess.* 8(1). https://doi.org/10.1186/s40643-021-00384-4

Ding, H., Li, Y., Lu, A., Jin, S., Quan, C., Wang, C., Wang, X., Zeng, C., Yan, Y., 2010. Photocatalytically improved azo dye reduction in a microbial fuel cell with rutile-cathode. *Bioresour. Technol.* 101(10), 3500–3505. https://doi.org/10.1016/j.biortech.2009.11.107

Dzurik, A.A., Kulkarni, T.S., Boland, B.K., 2018. *Water Resources Planning: Fundamentals for an Integrated Framework*, 4th ed. Rowman & Littlefield.

Edokpayi, J.N., Odiyo, J.O., Durowoju, O.S., 2017. Impact of wastewater on surface water quality in developing countries: A case study of South Africa. In *Water Quality*. InTech. https://doi.org/10.5772/66561

El Hakim, S., Chave, T., Nikitenko, S.I., 2021. Sonocatalytic degradation of EDTA in the presence of Ti and Ti@TiO$_2$ nanoparticles. *Ultrason. Sonochem.* 70, 105336. https://doi.org/10.1016/j.ultsonch.2020.105336

Feng, C.H., Li, F.B., Mai, H.J., Li, X.Z., 2010. Bio-electro-fenton process driven by microbial fuel cell for wastewater treatment. *Environ. Sci. Technol.* 44(5), 1875–1880. https://doi.org/10.1021/es9032925

Fu, J., Zhang, X., Li, H., Chen, B., Ye, S., Zhang, N., Yu, Z., Zheng, J., Chen, B., 2022. Enhancing electronic metal support interaction (EMSI) over Pt/TiO$_2$ for efficient catalytic wet air oxidation of phenol in wastewater. *J. Hazard. Mater.* 426, 128088. https://doi.org/10.1016/j.jhazmat.2021.128088

Fu, L., You, S.J., Zhang, G.Q., Yang, F.L., Fang, X.H., 2010. Degradation of azo dyes using in-situ Fenton reaction incorporated into H$_2$O$_2$-producing microbial fuel cell. *Chem. Eng. J.* 160(1), 164–169. https://doi.org/10.1016/j.cej.2010.03.032

Gai, H., Liu, X., Feng, B., Gai, C., Huang, T., Xiao, M., Song, H., 2021. An alternative scheme of biological removal of ammonia nitrogen from wastewater–highly dispersed Ru cluster @mesoporous TiO2 for the catalytic wet air oxidation of low-concentration ammonia. *Chem. Eng. J.* 407, 127082. https://doi.org/10.1016/j.cej.2020.127082

Gautam, P., Kumar, S., Lokhandwala, S., 2019. Advanced oxidation processes for treatment of leachate from hazardous waste landfill: A critical review. *J. Clean. Prod.* 237, 117639. https://doi.org/10.1016/j.jclepro.2019.117639

Gernaey, K.V., Van Loosdrecht, M.C.M., Henze, M., Lind, M., Jørgensen, S.B., 2004. Activated sludge wastewater treatment plant modelling and simulation: State of the art. *Environ. Modell. Softw.* 19(9), 763–783. https://doi.org/10.1016/j.envsoft.2003.03.005

Goel, R.K., Flora, J.R.V., Chen, J.P., 2005. Flow equalization and neutralization. In *Physicochemical Treatment Processes*. Humana Press, pp. 21–45. https://doi.org/10.1385/1-59259-820-x:021

Gong, W.-J., Li, F., Xi, D.-L., 2008. Oxidation of industrial dyeing wastewater by supercritical water oxidation in transpiring-wall reactor. *Water Environ. Res.* 80(2), 186–192. https://doi.org/10.2175/106143007x221067

Greyson, J.C., 2020. *Carbon, Nitrogen, and Sulfur Pollutants and Their Determination in Air and Water*. CRC Press.

Güngören Madenoğlu, T., Üremek Cengiz, N., Sağlam, M., Yüksel, M., Ballice, L., 2016. Catalytic gasification of mannose for hydrogen production in near- and super-critical water. *J. Supercrit. Fluids* 107, 153–162. https://doi.org/10.1016/j.supflu.2015.09.003

Gupta, P., Verma, N., 2021. Evaluation of degradation and mineralization of glyphosate pollutant in wastewater using catalytic wet air oxidation over Fe-dispersed carbon nanofibrous beads. *Chem. Eng. J.* 417, 128029. https://doi.org/10.1016/j.cej.2020.128029

Gürel, M., Pehlivanoglu-Mantas, E., 2010. Water quality requirements. In Krantzber, G. (ed.), *Advances in Water Quality Control*. Scientific Research Publishing, pp. 1–25.

Hassan, H., Hameed, B.H., 2011. Fe-clay as effective heterogeneous Fenton catalyst for the decolorization of reactive blue 4. *Chem. Eng. J.* 171(3), 912–918. https://doi.org/10.1016/j.cej.2011.04.040

Hussain, S., Aneggi, E., Trovarelli, A., Goi, D., 2022. Removal of organics from landfill leachate by heterogeneous fenton-like oxidation over copper-based catalyst. *Catalysts* 12(3), 1–17. https://doi.org/10.3390/catal12030338

Jeong, H., Kim, H., Jang, T., 2016. Irrigation water quality standards for indirect wastewater reuse in agriculture: A contribution toward sustainablewastewater reuse in South Korea. *Water (Switzerland)* 8(4). https://doi.org/10.3390/w8040169

Kesari, K.K., Soni, R., Jamal, Q.M.S., Tripathi, P., Lal, J.A., Jha, N.K., Siddiqui, M.H., Kumar, P., Tripathi, V., Ruokolainen, J., 2021. Wastewater treatment and reuse: A review of its applications and health implications. *Water Air Soil Pollut.* 232(5). https://doi.org/10.1007/s11270-021-05154-8

Khouni, I., Marrot, B., Moulin, P., Ben Amar, R., 2011. Decolourization of the reconstituted textile effluent by different process treatments: Enzymatic catalysis, coagulation/flocculation and nanofiltration processes. *Desalination* 268(1–3), 27–37. https://doi.org/10.1016/j.desal.2010.09.046

Kim, K.H., Ihm, S.K., 2011. Heterogeneous catalytic wet air oxidation of refractory organic pollutants in industrial wastewaters: A review. *J. Hazard. Mater.* 186(1), 16–34. https://doi.org/10.1016/j.jhazmat.2010.11.011

Kılıç, A., Uysal, Y., Çınar, Ö., 2011. Laboratuvar Ölçekli Bir Mikrobiyal Yakıt Hücresinde Sentetik Atıksudan elektrik Üretimi. *Pamukkale Üniversitesi Mühendislik Bilim. Derg.* 17(1), 43–49.

Kumar, J., Bansal, A., 2013. Photocatalysis by nanoparticles of titanium dioxide for drinking water purification: A conceptual and state-of-art review. *Mater. Sci. Forum* 764, 130–150. https://doi.org/10.4028/www.scientific.net/MSF.764.130

Kurniawan, S.B., Ahmad, A., Said, N.S.M., Imron, M.F., Abdullah, S.R.S., Othman, A.R., Purwanti, I.F., Hasan, H.A., 2021. Macrophytes as wastewater treatment agents: Nutrient uptake and potential of produced biomass utilization toward circular economy initiatives. *Sci. Total Environ.* 790, 148219. https://doi.org/10.1016/j.scitotenv.2021.148219

Kurniawan, T.A., Chan, G.Y.S., Lo, W.H., Babel, S., 2006. Physico-chemical treatment techniques for wastewater laden with heavy metals. *Chem. Eng. J.* 118(1–2), 83–98. https://doi.org/10.1016/j.cej.2006.01.015

Kurniawan, T.A., Mengting, Z., Fu, D., Yeap, S.K., Othman, M.H.D., Avtar, R., Ouyang, T., 2020. Functionalizing TiO_2 with graphene oxide for enhancing photocatalytic degradation of methylene blue (MB) in contaminated wastewater. *J. Environ. Manag.* 270, 110871. https://doi.org/10.1016/j.jenvman.2020.110871

Levec, J., Pintar, A., 2007. Catalytic wet-air oxidation processes: A review. *Catal. Today* 124(3–4), 172–184. https://doi.org/10.1016/j.cattod.2007.03.035

Li, D., Liu, S., 2019. *Water Quality Monitoring and Management*. Elsevier. https://doi.org/10.1016/C2016-0-00573-9

Li, Y., Lu, A., Ding, H., Wang, X., Wang, C., Zeng, C., Yan, Y., 2010. Microbial fuel cells using natural pyrrhotite as the cathodic heterogeneous Fenton catalyst towards the degradation of biorefractory organics in landfill leachate. *Electrochem. Commun.* 12(7), 944–947. https://doi.org/10.1016/j.elecom.2010.04.027

Liu, D., Liptak, B.G., 1999. *Environmental Engineers' Handbook*, 2nd ed. CRC Press.

Liu, D.H.F., Liptak, B.G., 1997. *Environmental Engineers' Handbook*. CRC Press.

Liu, X., Sun, C., Chen, L., Yang, H., Ming, Z., Bai, Y., Feng, S., Yang, S.T., 2018. Decoloration of methylene blue by heterogeneous Fenton-like oxidation on Fe_3O_4/SiO_2/C nanospheres in neutral environment. *Mater. Chem. Phys.* 213, 231–238. https://doi.org/10.1016/j.matchemphys.2018.04.032

Malakootian, M., Nasiri, A., Amiri Gharaghani, M., 2020. Photocatalytic degradation of ciprofloxacin antibiotic by TiO_2 nanoparticles immobilized on a glass plate. *Chem. Eng. Commun.* 207(1), 56–72. https://doi.org/10.1080/00986445.2019.1573168

Malik, P.K., Saha, S.K., 2003. Oxidation of direct dyes with hydrogen peroxide using ferrous ion as catalyst. *Sep. Purif. Technol.* 31(3), 241–250. https://doi.org/10.1016/S1383-5866(02)00200-9

Manahan, S.E., 2010. *Fundamentals of Environmental Chemistry*, 3rd ed. CRC Press. https://doi.org/10.1201/9781420056716

Mantzavinos, D., 2007. Basic unit operations in wastewater treatment. In *Utilization of By-Products and Treatment of Waste in the Food Industry.* Springer, pp. 31–51. https://doi.org/10.1007/978-0-387-35766-9_3

Matko, T., Fawcett, N., Sharp, A., Stephenson, T., 1996. Recent progress in the numerical modelling of wastewater sedimentation tanks. *Process Saf. Environ. Prot.* 74(4), 245–258. https://doi.org/10.1205/095758296528590

Moradi, N., Amin, M.M., Fatehizadeh, A., Ghasemi, Z., 2018. Degradation of UV-filter Benzophenon-3 in aqueous solution using TiO_2 coated on quartz tubes. *J. Environ. Heal. Sci. Eng.* 16(2), 213–228. https://doi.org/10.1007/s40201-018-0309-3

Naik, S., Jujjavarappu, S.E., 2020. Simultaneous bioelectricity generation from cost-effective MFC and water treatment using various wastewater samples. *Environ. Sci. Pollut. Res. Int.* 27(22), 27383–27393. https://doi.org/10.1007/s11356-019-06221-8

Nemerow, N.L. (Ed.), 2007. Removal of suspended solids. In *Industrial Waste Treatment.* Elsevier, pp. 53–77. https://doi.org/10.1016/B978-012372493-9/50040-5

Nemerow, N.L., Agardy, F.J., Sullivan, P.J., Salvato, J.A., 2009. *Environmental Engineering.* Wiley. https://doi.org/10.1002/9780470432808

Ni, B.J., Yu, H.Q., 2012. Microbial products of activated sludge in biological wastewater treatment systems: A critical review. *Crit. Rev. Environ. Sci. Technol.* 42(2), 187–223. https://doi.org/10.1080/10643389.2010.507696

Nidheesh, P.V., Gandhimathi, R., Ramesh, S.T., 2013. Degradation of dyes from aqueous solution by Fenton processes: A review. *Environ. Sci. Pollut. Res. Int.* 20(4), 2099–2132. https://doi.org/10.1007/s11356-012-1385-z

O'Shea, K.E., Dionysiou, D.D., 2012. Advanced oxidation processes for water treatment. *J. Phys. Chem. Lett.* 3(15), 2112–2113. https://doi.org/10.1021/jz300929x

Ofori, S., Puškáčová, A., Růžičková, I., Wanner, J., 2021. Treated wastewater reuse for irrigation: Pros and cons. *Sci. Total Environ.* 760. https://doi.org/10.1016/j.scitotenv.2020.144026

Ovejero, G., Sotelo, J.L., Rodríguez, A., Vallet, A., García, J., 2011. Wet air oxidation and catalytic wet air oxidation for dyes degradation. *Environ. Sci. Pollut. Res. Int.* 18(9), 1518–1526. https://doi.org/10.1007/s11356-011-0504-6

Pang, Y.L., Abdullah, A.Z., Bhatia, S., 2011. Review on sonochemical methods in the presence of catalysts and chemical additives for treatment of organic pollutants in wastewater. *Desalination* 277(1–3), 1–14. https://doi.org/10.1016/j.desal.2011.04.049

Park, H., Park, Y., Kim, W., Choi, W., 2013. Surface modification of TiO_2 photocatalyst for environmental applications. *J. Photochem. Photobiol. C Photochem. Rev.* 15, 1–20. https://doi.org/10.1016/j.jphotochemrev.2012.10.001

Patnaik, P., 2017. *Handbook of Environmental Analysis*, 3rd ed. Boca Raton, Taylor & Francis, CRC Press. https://doi.org/10.1201/9781315151946

Pérez-Lucas, G., El Aatik, A., Aliste, M., Hernández, V., Fenoll, J., Navarro, S., 2022. Reclamation of aqueous waste solutions polluted with pharmaceutical and pesticide residues by biological-photocatalytic (solar) coupling in situ for agricultural reuse. *Chem. Eng. J.* 448. https://doi.org/10.1016/j.cej.2022.137616

Phocaides, A., F.A.O., 2007. *Handbook on Pressurized Irrigation Techniques.* Food & Agriculture Org.

Pipi, A.R.F., Sirés, I., De Andrade, A.R., Brillas, E., 2014. Application of electrochemical advanced oxidation processes to the mineralization of the herbicide diuron. *Chemosphere* 109, 49–55. https://doi.org/10.1016/j.chemosphere.2014.03.006

Pradhan, S.R., Colmenares-Quintero, R.F., Quintero, J.C.C., 2019. Designing micro-flowreactors for photocatalysis using sonochemistry: A systematic review article. *Molecules* 24(18). https://doi.org/10.3390/molecules24183315

Qazi, U.Y., Iftikhar, R., Ikhlaq, A., Riaz, I., Jaleel, R., Nusrat, R., Javaid, R., 2022. Application of Fe-RGO for the removal of dyes by catalytic ozonation process. *Environ. Sci. Pollut. Res. Int.*, 89485–89497. https://doi.org/10.1007/s11356-022-21879-3

Rahim Pouran, S., Abdul Aziz, A.R., Wan Daud, W.M.A., 2015. Review on the main advances in photo-Fenton oxidation system for recalcitrant wastewaters. *J. Ind. Eng. Chem.* 21, 53–69. https://doi.org/10.1016/j.jiec.2014.05.005

Rashed Nageeb, M., 2013. Adsorption technique for the removal of organic pollutants from water and wastewater. In *Organic Pollutants - Monitoring, Risk and Treatment*. InTech. https://doi.org/10.5772/54048

Rauf, M.A., Meetani, M.A., Hisaindee, S., 2011. An overview on the photocatalytic degradation of azo dyes in the presence of TiO_2 doped with selective transition metals. *Desalination* 276(1–3), 13–27. https://doi.org/10.1016/j.desal.2011.03.071

Roy, S., Saroha, A.K., 2014. Ceria promoted γ-Al_2O_3 supported platinum catalyst for catalytic wet air oxidation of oxalic acid: Kinetics and catalyst deactivation. *RSC Adv.* 4(100), 56838–56847. https://doi.org/10.1039/c4ra06529h

Samer, M., 2015. Biological and chemical wastewater treatment processes. In *Wastewater Treatment Engineering*. InTech. https://doi.org/10.5772/61250

Sang, Y., Liu, H., Umar, A., 2015. Photocatalysis from UV/Vis to near-infrared light: Towards full solar-light spectrum activity. *ChemCatChem* 7(4), 559–573. https://doi.org/10.1002/cctc.201402812

Singh, L., Rekha, P., Chand, S., 2016. Cu-impregnated zeolite Y as highly active and stable heterogeneous Fenton-like catalyst for degradation of Congo red dye. *Sep. Purif. Technol.* 170, 321–336. https://doi.org/10.1016/j.seppur.2016.06.059

Song, Y., Feng, S., Qin, W., Li, J., Guan, C., Zhou, Y., Gao, Y., Zhang, Z., Jiang, J., 2022. Formation mechanism and control strategies of N-nitrosodimethylamine (NDMA) formation during ozonation. *Sci. Total Environ.* 823, 153679. https://doi.org/10.1016/j.scitotenv.2022.153679

Sonune, A., Ghate, R., 2004. Developments in wastewater treatment methods. *Desalination* 167, 55–63. https://doi.org/10.1016/j.desal.2004.06.113

Spellman, F.R., 2013. *Handbook of Water and Wastewater Treatment Plant Operations*, 3rd ed. CRC Press. https://doi.org/10.1201/b15579

Stoddard, A., Harcum, J.B., Bastian, R.K., Simpson, J.T., Pagenkopf, J.R., 2003. *Municipal Wastewater Treatment: Evaluating Improvements in National Water Quality*. John Wiley & Sons.

Tebbutt, T.H.Y., 1998. Water — A precious natural resource. In *Principles of Water Quality Control*. Elsevier, pp. 1–11. https://doi.org/10.1016/B978-075063658-2/50002-1

Tomar, M., 1999. *Quality Assessment of Water and Wastewater*, 1st ed. CRC Press. https://doi.org/10.1201/9781482264470

Tungler, A., Szabados, E., Hosseini, A.M., 2015. Wet air oxidation of aqueous wastes. In *Wastewater Treatment Engineering*. InTech. https://doi.org/10.5772/60935

Uçar, D., Toprak, D., 2014. Several methods to separate anode and cathode chambers in microbial fuel cells. *Afyon Kocatepe Univ. J. Sci. Eng.* 14(1), 1–6. https://doi.org/10.5578/fmbd.6616

Valange, S., Védrine, J.C., 2018. General and prospective views on oxidation reactions in heterogeneous catalysis. *Catalysts* 8(10). https://doi.org/10.3390/catal8100483

Vesilind, P.A., DiStefano, T.D., 2006. *Controlling Environmental Pollution: An Introduction to the Technologies, History and Ethics*. DEStech Publications.

Vesilind, P.A., Peirce, J.J., Weiner, R.F., 2013. *Environmental Engineering.* Butterworth-Heinemann.

Weiner, R., Matthews, R., 2003. *Environmental Engineering.* Elsevier. https://doi.org/10 .1016/B978-0-7506-7294-8.X5000-3

Wu, C.H., Kuo, C.Y., Chang, C.L., 2008. Decolorization of C.I. Reactive Red 2 by catalytic ozonation processes. *J. Hazard. Mater.* 153(3), 1052–1058. https://doi.org/10 .1016/j.jhazmat.2007.09.058

Yenkie, K.M., Pimentel, J., Orosz, Á., Cabezas, H., Friedler, F., 2021. The P-graph approach for systematic synthesis of wastewater treatment networks. *AIChE J.* 67(7), 1–10. https://doi.org/10.1002/aic.17253

Zhao, J., Wei, Y., Liu, Z.J., Zhang, L., Cui, Q., Wang, H.Y., 2022. Study on heterogeneous catalytic wet air oxidation process of high concentration MDEA-containing wastewater. *Chem. Eng. Process. Process Intensif.* 171, 108744. https://doi.org/10.1016/j .cep.2021.108744

Zhao, X., Jin, H., 2020. Correlation for self-diffusion coefficients of H_2, CH_4, CO, O_2 and CO_2 in supercritical water from molecular dynamics simulation. *Appl. Therm. Eng.* 171. https://doi.org/10.1016/j.applthermaleng.2020.114941

Zhou, M., He, J., 2007. Degradation of azo dye by three clean advanced oxidation processes: Wet oxidation, electrochemical oxidation and wet electrochemical oxidation-A comparative study. *Electrochim. Acta* 53(4), 1902–1910. https://doi.org/10 .1016/j.electacta.2007.08.056

Zhu, X., Ni, J., 2009. Simultaneous processes of electricity generation and p-nitrophenol degradation in a microbial fuel cell. *Electrochem. Commun.* 11(2), 274–277. https:// doi.org/10.1016/j.elecom.2008.11.023

6 Bridging Corporate Social Responsibility to Green Chemistry with Reference to Steel Industry

Meghna Sharma, Aastha Anand, and Sanjay K. Sharma

CONTENTS

6.1 INTRODUCTION

An upward trend is found in the chemical industries in India to meet the desired demand nationally and globally, and the chemical industries in India are flourishing. With the continuous rise in these industries, one should also look at the impact that these industries have on consumers as well as the environment. When the environmental aspect is concerned, it is where green chemistry comes into the

DOI: 10.1201/9781003301769-6

picture. Often, the products we consume are a combination of several chemical activities, and these activities should be traced down to know the impact that they are having on the environment. With the ongoing scenario that the world is facing such a crucial time where natural resources are on the verge of extinction, green chemistry can be proven to have a drastic positive impact on the concepts that companies follow while making a product.

The term sustainability is unavoidable given the state of affairs. Sustainability means that we use the resources in such a way that it is least harmful to the eco-system, while preserving the same for future generations. It is using the resources judiciously to preserve the environment and conserve finite supplies.

Sustainability is the ability to meet the needs of the present without com-promising the ability of the future generation to meet their own needs (UN World Commission on Environment and Development). On April 1, 2014, India became the first country to legally mandate corporate social responsibility (CSR). According to Section 135 of India's Companies Act, companies are asked to set aside 2% of their average net profit for social responsibility. If one needs to imple-ment corporate social responsibility, then they need to fulfill two components that are creation of real benefit for the community and building a strong relationship with their collaborators that could be instrumental or strategic.

If an organization follows these two components, then it will have the oppor-tunity to convert corporate social responsibility to corporate social innovation (CSI). Innovation is a core engine for market growth and thus plays a vital role in improving and maintaining the competitive edge of businesses and is a source for improving the standard of living of the people. Social innovation is known as a process that develops as well as deploys effective solutions for the challenging social and environmental issues that play an important role in supporting social progress.

This study is inclined toward corporate social innovation and how certain organizations are giving rise to corporate social innovation by converting corpo-rate social responsibility into corporate social innovation. Necessity is known to be the mother of all innovation, as a result many corporate social responsibility activities are creating a piece that is unique and further empowers society. Strong relationships and a sense of belongingness are the outcomes of this innovation-driven social response.

Corporate social innovation engages itself in society-driven R&D to build a true partnership between companies, agencies, and NGOs. While corporate social responsibility is providing social and economic services to the needy (Carmen stoian, 2016), corporate social innovation with their partners tends to create something new and something innovative to provide more sustainable solu-tions for the people who need it. Social innovation aims to open a whole new mar-ket or segment that involves innovation in the entire business model. Tilt (2016) in her paper states that examining without contextualization of CSR and its report-ing activities may reinforce inaccurate understandings that are based on research results in the developed world. For any corporate, innovation is what helps them in sustaining the red ocean.

The need for social innovation arises whenever there is any gap between expected and actual reality in society. For any innovation to be called a social innovation, there are certain parameters that should be fulfilled: first, the innovation should be done strictly to bring change to the society, and second that there needs to be social value creation through this innovation. Social innovation tends to differentiate themselves from others by the dignity of their intention toward goals and mission with respect to society. Saluja and Kapoor (2017) in their study talked about the evolution of corporate social responsibility, they talked about how CSR became a sustainable initiative from responsive activities. Social innovation involves a societal group that establishes a supportive relationship among its member to create social vision and to create solutions for the problems in society. Therefore, social innovators serve as an agent of change for society, by taking advantage of the lost resources and better processes as well as creating innovative strategies and solutions to rebuild the society. Ghosh (2014) in her study investigates how private sector corporations can engage in corporate social responsibility activities as expressed in their respective public domain corporate documents with absolute benefits as their parameter

In general terms, social innovation is a new form of innovation that helps in solving social and economic problems and further helps in social change (Carroll, 1984). Social innovation has seen various obstacles like deviations from the mission, red ocean, i.e., cut-throat competition, limitation of resources, lack of governmental intervention, and so on. This present paper focuses on highlighting the concept of social innovation and social responsibility, drivers of social innovation, and social responsibility, as well as exploring various innovation initiatives taken by organizations. Rubalcaba (2016) studied social innovation research: an emerging area of innovation studies reveals how social innovation literature came into existence after the year 2002. This study talks about the content scope of social innovation research by using bibliometric analysis and they further explore the relevance of the study. He talks about how social technology and value creation together sum up the need for innovation studies.

Verma and Singh (2016) in their study have shown how corporate social responsibility is an integral part of organization's competitive strategy. They also talk about how organizations use their corporate social responsibility as a signal to win over loyal consumers.

Green chemistry is how a particular product should be designed chemically to reduce its harmful impact on the environment. It is concerned with the production of the product to its use and thereafter its ultimate disposal. It is the art of constructing an innovative way that can be fruitful for companies, the environment, and the people. Green chemistry can be in the form of reduction of waste while producing a product, increasing energy effectiveness, and carefully doing waste control, and so on. It can be in any structure like alterations to existing technology, replacement of raw material with a more environmentally friendly alternative, alteration, and redesigning of the products, all these changes are done by keeping one thing in mind that is of utmost importance that we have to be sustainable in each and every way that is possible to preserve our mother earth.

Evolution of an economy into an environmentally friendly nation depends on several factors and metrics (Sheldon, 2017). Minimization of trash and maximization of resource efficacy are crucial factors in determining greenness. Certain parameters should be developed that can evaluate the environmental influence of raw materials and waste. Another standpoint that can be constructed is the shift or evolution of an unsustainable economy that uses fossil fuels to a sustainable economy that uses fuels that are not harmful to the environment. All this comprises green chemistry. The thoughts of green chemistry and sustainable chemistry have meanings that are identical as well as contrasting at the same time. The attributes green and sustainable have varying implications, particularly because of the vicinity wherein they are used. Currently, green/sustainable chemistry is discovered in all nations throughout the world playing a critical position in investigation, collaboration, initiatives, networks, teaching, and university curricula.

The green chemistry concept was first developed by Joseph Breen, Paul Anastas, and Tracy Williamson in 1998: "The invention, design, and application of chemical products and processes to reduce or to eliminate the use and generation of hazardous substances".

Through this chapter, we aim to find out:

a) To understand the relationship between green chemistry and corporate social responsibility.
b) To understand CSR as a provenance for CSI.
   This chapter will explore the meaning of green chemistry and how CSR activities can help a company to see a clearer picture of green chemistry, it will explore how CSR initiatives will eventually have an impact on green chemistry. Green chemistry is an emerging topic nowadays. Many organizations are practicing green chemistry to have a positive impact on the organization itself and its employees, stakeholders, and ultimately the environment.

6.2 METHODOLOGY

This chapter explores the conceptual understanding of corporate social responsibility and corporate social innovation and green chemistry. A variety of academic papers have been consulted to obtain a proper definition of the research subject: "To understand the relationship between green chemistry and corporate social responsibility" and "Corporate social responsibility as a provenance for corporate social innovation".

A secondary study has been conducted in this chapter. For writing this chapter, various research papers, articles, and journals have been consulted in order to get the appropriate insights into the topic.

In this chapter, we determine what is known about the performance of CSR initiatives, and we develop a path forward to open the opportunity for CSR to become a source of social innovation. Thereafter, we dig into recent studies to identify the drivers of CSI. From these analyses, we determine that although the literature has advanced in specific ways, it continues to provide inadequate insight into CSR as a genesis of CSI. We then put forth a revised research agenda, that can overcome these limitations, enabling researchers to determine how effectively CSR initiatives can lead to social innovation.

We also dig deeper into the concept of green chemistry with the help of various research articles and attempted to understand the relationship between corporate social responsibility and green chemistry and how CSR initiatives can give rise to green chemistry activities.

6.3 THE RELATIONSHIP BETWEEN GREEN CHEMISTRY AND CORPORATE SOCIAL RESPONSIBILITY

The concept of CSR is transforming, CSR is more inclined toward the chemical industry, and it is time that we act. A civilization wants all the gains that they can get from the chemical industry without having to face any of the negative consequences that this industry has on the environment and on the health of the consumers and the workers. This is where green chemistry comes into force. It is the need of the society that an organization should produce in such a way that is least harmful to the environment while most advantageous for its audience that the organization must capture. On the other hand, companies are engaging in such activities that fulfill the criteria while making profits and maintaining a good corporate image by following CSR initiatives. This gives furthermore reason for an organization to practice green chemistry.

While following CSR activities, a company integrates the concept of sustainability and green chemistry into its business activities so that it can have a positive influence on a company's revenue. Green chemistry can be embraced by a company as a form of social innovation, which can help to create a more sustainable product and service that a company can offer to its targeted audience.

The integration of CSR is taking benefit of the innovative prospects, which are growing and intend to safeguard commitment with consumers. There should be a modern understanding of thinking, and new approaches to working collectively through the values chain, which needs the sense of new technologies the revolution adapted viewpoint presents significant chances for a unique competitive advantage. This is often extremely valuable over competitor firms by opening to prospective new buyers, with distinction in a congested market, and increasing trades and returns through increased manufacturing capacity and reduced costs.

If green chemistry observes the route to omnipresence, which sustainability has inside the chemicals business, the expected market capacity will become a reality with a more sustainable presence.

The chemicals industry has shown considerable progressive moves and created numerous excellent accomplishments providing socio-economic improvement. However, the industry has been confronting rather limited environmental and social challenges, which involve it exploring the latest routes in development, construction, and utilization to ensure sustainable development.

Investigation reveals that garbage avoidance is better and costs a smaller amount than the treatment or dumping of waste once it has been produced. Furthermore, the more excessive the garbage companies generate, the greater the expense that they must take on to trade with the environmental, social, and public health outcomes.

Implementing a green chemistry methodology in the construction of goods and services also inculcates that the organizations also incorporate the point that companies take on their social responsibility. CSR can be shown in these classifications:

- CSR for personnel: Guarantee a secure operating atmosphere, encourage impartiality for females and males on all stages, foster working prospects, and safeguard confidentiality and security for employees.
- CSR for buyers and customers: Offer superior-value goods produced from secure, non-toxic substances at an acceptable cost, and perform investigation and commodity advancement to mercenaries of advanced and valuable commodities.
- CSR for merchants (e.g., providers of sustainable materials): Offer learning for suppliers to produce better, safe, and sound supplies. CSR activities should be practiced at each level for it to become an important activity. Everyone should perform sustainable practices in his or her capacity because every effort counts. Each ounce of endeavor that goes into transforming the world into a sustainable economy matters.
- CSR for the people: Implement programs to improve the community and the region, aiming at where manufacturing works or suppliers are situated. Educating the community that they should be aware of the CSR activities that the organizations are performing, and they should promote such businesses.
- CSR for the environment and well-being: Purchase resources that are harmless for the environment and society, prevent damaging flora and fauna, guarantee that the production process is unpolluted; ensure transparency when communicating environmental information; support environmentally friendly packaging; and develop products and procedures that are as environmentally friendly as possible, focusing on voluntary participation in the GC process and considering GC as part of the company's CSR policy.

For a nation that wants sustainability to be its priority the buyers should have access to sustainable products, and for sustainable goods, we need green raw materials, sustainable and responsible production, and green creation. When we are cognizant that green chemistry helps us to protect our health and with the help of green chemistry, we can produce sustainable green products then both the consumers and manufacturers will willingly take a step forward to promote green chemistry. Green chemistry is an unavoidable shift that is being undertaken on a universal scale, which prevents us to steer clear of previous misjudgments such as the invention and use of harmful chemicals and substances or the production and use of chemicals in the production or composition of a product, even very harmful chemicals are still used for the disposal of waste. The execution of CSR initiatives involves numerous endeavors and the deliberate devotion of the managers and workers of the enterprise, as well as support from other businesses in the industry. These deliberate CSR pursuits bring companies not only gains but also faith from the marketplace and the society.

Since the early nineties, societal actions have transformed green chemistry and have given changes in manufacturing situations and sustainable practices with developments in ecological influence and understanding of corporations and populace. Paul Anastas and John Warner, in the 1990s, suggested the 12 standards of green chemistry that are established on the decrease or non-use of harmful/lethal cleaners in compound procedures and evaluate as well as the non-generation of remains from these practices (Marco et al., 2018).

6.4   GREEN CHEMISTRY PRACTICES IN SWEDISH STEEL INDUSTRY—INDIA

A new Hydrogen Breakthrough Ironmaking Technology (HYBRIT) proposal, Swedish Steel India has carried up green chemistry/production program that aims to substitute coking coke with hydrogen in the steel production method. The steel industry is one of the top carbon dioxide-producing businesses, with approximately reporting for 7% of overall releases worldwide. HYBRIT (Hydrogen Breakthrough Ironmaking Technology) intends to substitute coal with hydrogen in the steelmaking procedure, which will change the current blast furnace procedure. Presently, coke is applied to transform the iron ore into iron, but in the current procedure, the coke will be swapped by hydrogen gas, which is generated by fossil-free power supplies. The by-product will be water that can be regained for the manufacturing of hydrogen gas. As per the corporation, the decrease in reactions in ironmaking represents approximately 85–90% of the total carbon dioxide discharges in the ore-based steelmaking value chain. In the case of HYBRIT, iron metal is manufactured by using hydrogen gas as the prime reductant.

6.5   CSR AS A PROVENANCE TO CSI

Paunescu (2014) discusses the trends that are most popular in social innovation and how corporate social responsibility has given rise to social innovation. This paper talks about how social innovation in the field of research has gathered popularity in the past decade among researchers. This research was done with the motive of helping scholars to advance their research in the field of social innovation. Simon Slavin (1993) in her article on social innovation tells us that doing good for society is what is social innovation all about. She further stated that there are two broad categories of people one who sees social innovation as a mode for developing products or services that serve the betterment of the society and others who believe that business should not interfere in social innovation because business is all about profit and not people. Martinez et al. (2020)) in their paper examine the concept of social innovation in business and how it is different from social innovation that is been carried by civil society. They found that social innovation can be best explained as a process that is been carried by humans who fulfill moral, creativity, and path dependencies of individuals. Social innovations are criticized by a social purpose and goal, and their importance is important for all participants to share interest (economic and social). The social context in question has influenced the views of innovation not only as a weapon and source of economic growth and competition, but as a possible tool to achieve social goals and social stability in society (Gabriela, 2012).

Social innovation is the creation and implementation of effective solutions to social and environmental challenges, often systematic, that promote social progress.

> Social innovation is not the prerogative or privilege of any organizational form or legal structure. Solutions often require the active collaboration of constituents across government, business, and the nonprofit world.
>
> **Soule, Malhotra, Clavier**

6.5.1 CORPORATE SOCIAL RESPONSIBILITY

CSR is all about the humanitarian intent done for the society by an organization to help the society by providing them with money or certain personnel (if required). To conduct these CSR activities, companies associate themselves with certain NGOs those will carry out these social services on behalf of these organizations. CSR initiatives are done only for those communities who are in need of it.

6.5.2 CORPORATE SOCIAL INNOVATION

Corporate social innovation is known as a strategy-driven intent by organizations for the society. Corporate social innovation engages itself in societally relevant research and development and if required also contributes its corporate assets. Unlike corporate social responsibility, corporate social innovation comes from partnerships involving other organizations, NGOs, and government agencies. Whenever social innovation is done, it is made keeping in mind that this product/ service will create an innovative outcome that will further provide solutions to the problems of society (Table 6.1).

6.5.2.1 Elements of Corporate Social Innovation

6.5.2.1.1 Purpose

For any organization to start social innovation, they need to must have a persuasive vision for the betterment of the world and a mission that holds strong moral values. A vision illustrates a company's desired future. The scenario for businesses to develop social change is vision, mission, and values (Figure 6.1).

TABLE 6.1
Difference between CSR and CSI

Corporate Social Responsibility	Corporate Social Innovation
CSR is a humanitarian intent of companies toward society. It aims to manage all internal business practices in a way that considers the interest of all stakeholders and the environment.	CSI is a strategy-driven intent by companies. CSI programs are outside of a company's ordinary business practices and are largely developmental in nature.
CSR involves in the contribution of money and manpower to society.	CSI engages itself in societally relevant research and development and contributes corporate assets.
CSR is a continuous practice that improves the quality of life for businesses, stakeholders, and employees while improving company's image and making profits.	CSI projects are developmental in nature. They are meant to develop and uplift communities. It is done by companies to benefit society without the intention of generating profits.
CSR is providing social service to those who need it.	CSI tends to create something innovative by providing solutions to the needs and problems of society.

FIGURE 6.1   Elements of corporate social innovation.

6.5.2.1.2 Intent
Strategic intent is what is required by organizations to work on social innovation. It means innovation by helping companies by identifying an opportunity and by developing a value proposal for the organization and their customers. Strategic intent translates purpose into direction for innovation by helping companies construct a value for the organization.

6.5.2.1.3 Partner
Corporate social innovation is designed to match the core competencies of a company to the pressing needs of society and produce successful innovations. In order to do so, social innovation needs a diverse range of interests, perspectives, and skills. There are only a few firms with an adequate mix of staff, resources, and capabilities required to innovate themselves in this area in order to become competitive various organizations that tend to tie up with different organizations, agencies, and NGOs for fulfilling the need for matching competencies.

6.5.2.1.4 Process
Innovation is unleashing the boundaries of R&D and now emphasizing on user-driven organization. Business innovation goes beyond the limits of R&D

by emphasizing innovation and co-creation that are driven by users. With respect to corporate social innovation, this presents the company with new opportunities and challenges. Firstly, it requires the promotion of communities and populations in need from outside their traditional customer base. Secondly, by co-creating partners' ecosystems involving different views on the issues involved and different ideas on the best way to deal with them. Thirdly, a constant adaptation of design and implementation to the "local" circumstances is needed.

6.5.2.1.5   Result

The benefits and returns on corporate social innovation raise questions on whether in helping societies and preserving the environment, social innovations can (and should) provide a value for business. There are several "purists" in academia and practice who emphasize "never anticipate or organize social entrepreneurs or innovators to create significant profits" and many who argue that social innovation returns should at best cover expenditure and not make profits for investors. However, recent research has shown that eco- and social innovation-based products and services can bring value to society and business. In the formulation of "shared value", Michael Porter and Mark Kramer rewrote early framings of "mixed value" and "socio-commercial business models". There is no doubt that the change in value creation opens new avenues. It immediately unites the interest of a company in creating new markets at the base of the pyramid by expanding its offering to ethically (or socially responsible) consumers.

Walter Isaacson wrote a biography about Steve Jobs, in which he discussed a significant period when Jobs returned to Apple after being fired years earlier. During this time, Jobs played a crucial role in saving Apple from impending failure. Jobs launched a new product line (iPod, iTune, iPhones, and iPads), as a result, Apple rises to the top of the industry in market capitalization, brand value, and customer appeal. He talks about how Jobs was once told by his colleague that "companies that last know how to re-invent themselves".

In overall terms, the main purpose of innovation is to improve the lives of people. Innovation is the key to making any kind of progress when it comes to running a business.

6.6   CORPORATE SOCIAL INNOVATION INITIATIVES

6.6.1   ITC AND e-CHOUPAL

e-Choupal is an ITC Limited enterprise project based in India that provides rural farmers with internet access. The goal is to educate and empower them, thereby improving the quality of agricultural products and the quality of life for farmers. The ITC's e-Choupal initiative is a strong example of a business model that creates broad societal value by co-creating rural markets with local communities. Realizing the various challenges faced by farmers, such as fragmented farmland, poor infrastructure, and the involvement of multiple intermediaries, ITC has built e-Choupal as a more efficient supply chain designed to deliver sustainable value to its customers worldwide. e-Choupal is an initiative initiated in the year 2000 by ITC Limited. With the aid of the internet, e-Choupal directly connects with rural farmers. e-Choupal means a place in Hindi where village meetings are held. e-Choupal is a virtual market where farmers can make proper transactions with buyers and reveal the actual value of their goods. e-Choupal is one of the main initiatives in rural India among all internet-based interventions.

6.6.2   TATA STRIVE: INITIATIVE BY TATA

Empowering youth with the right skills candefine the future of the country. Tata group as an industry leader is wellplaced to build a technology based scalable model for skilling, with a focus onquality of training, leading to employability and entrepreneurshipopportunities for youth.

(Mr S.Ramadorai; Chairman Tata STRIVE)

Tata STRIVE is an initiative started by Tata to train youth and make them employment ready. Two-thirds of India's 1.2 billion population is under 35 years, making the country one of the youngest in the world, from a demographic perspective. The economic advantage of having a population of such a large working age is clear. A significant percentage of that population is unqualified or under-skilled, thus capacity building is really a crucial national growth priority field. The outcome of India's skills project will be to turn this enormous human capital into successful workers and risky entrepreneurs. Developing a skilled workforce will also help India leverage a global opportunity by developing a skilled workforce. According to the 11th Five-Year Plan released by the Indian Government, the rest of the world will have a shortage of 55 million skilled workers in 2022, while India will have a surplus of 47 million skilled workers. Tata STRIVE is Tata Community Initiatives Trust's skill development initiative under the umbrella of Tata Trusts, addressing the pressing need to skill Indian youth for employment, entrepreneurship, and community entrepreneurship.

It enters neighborhoods, builds people's skills from financially challenged backgrounds, and acclimates them to changing working environments. The core philosophy is to create courses that would help to create and supply a trained workforce across the entire industrial spectrum, as well as develop entrepreneurial talent. Tata STRIVE actively builds partnerships with Tata companies, non-profits, non-Tata businesses, government agencies, foundations, trusts, and banks. Industrial Technical Institutes (ITIs) of India are a major part of the partnership model. Tata STRIVE works with credible non-governmental organizations, some of which may already have a working relationship with more than one of the Tata companies. This partnership is then built upon, through which Tata STRIVE brings expertise through information, technology, and process while leveraging the physical infrastructure. The algorithm used by Tata STRIVE is:

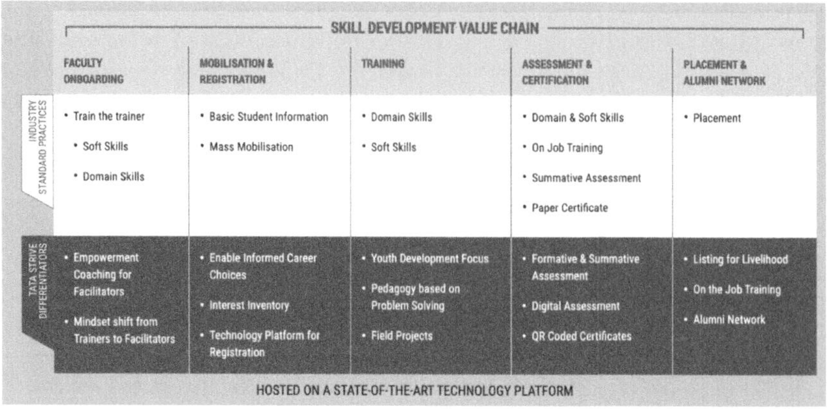

SKILL DEVELOPMENT VALUE CHAIN				
FACULTY ONBOARDING	**MOBILISATION & REGISTRATION**	**TRAINING**	**ASSESSMENT & CERTIFICATION**	**PLACEMENT & ALUMNI NETWORK**
• Train the trainer • Soft Skills • Domain Skills	• Basic Student Information • Mass Mobilisation	• Domain Skills • Soft Skills	• Domain & Soft Skills • On Job Training • Summative Assessment • Paper Certificate	• Placement
• Empowerment Coaching for Facilitators • Mindset shift from Trainers to Facilitators	• Enable Informed Career Choices • Interest Inventory • Technology Platform for Registration	• Youth Development Focus • Pedagogy based on Problem Solving • Field Projects	• Formative & Summative Assessment • Digital Assessment • QR Coded Certificates	• Listing for Livelihood • On the Job Training • Alumni Network

(left margin labels: INDUSTRY STANDARD PRACTICES; TATA STRIVE DIFFERENTIATORS)

HOSTED ON A STATE-OF-THE-ART TECHNOLOGY PLATFORM

FIGURE 6.2   Tata STRIVE—skill development value chain.

A) Ample opportunities: The core philosophy of Tata STRIVE is that no youth should be left behind they make sure that there are ample opportunities. They make sure that every decision, process, and investment is geared to mainstream the disadvantaged youth in the nation's economy.

B) Their methodology not only helps in nurturing technical skills but also helps with their soft skills. At Tata STRIVE, they closely monitor the market and then design and develop courses accordingly so that they can cater to the diverse demand of different industries (Figure 6.2).

Figure 6.2 depicts how the process followed by STRIVE is different from other practices followed by industries.

Once the course is decided, the next counselor meets the student and takes them through innovative audio–video materials. Realistic workplace photo lets them make educated choices. Tata STRIVE also invites students' parents to demonstrate the presentation of classroom sessions, tools, and equipment to be used and to educate them about the tasks to be achieved. Parents engage in the decision-making process, which inevitably leads to daily attendance and decreases the risk of dropouts.

6.7   BENEFIT BUSINESSES TEND TO ACQUIRE WHILE GREEN CHEMISTRY AND PURSUING SOCIAL INNOVATION

Sustainability and social innovation are known as the key to success in the market.

An accountable corporate image is hard to build but easy to lose. If an organization is making first mover advantage or becoming a leader for change, then it has a critical impact on the corporate image. Treating green chemistry and social innovation as a natural way of operating in an enterprise allows organizations to establish and maintain a corporate image. Attempts are made to address social challenges with the use of socially responsible solutions. For any organization, there is no escape from the requirement to consider the macroeconomic circumstances that will have a huge impact on your business and its future. Rapid shifts in the global climate have left businesses with no choice but to accept sustainability. Social innovativeness is not an option in this context; adaptation to the changing circumstances under which we live has become a requirement. Governments are introducing new legislation to encourage responsible business and prevent goods and processes that obstruct the road to improving our environmental footprint. Social innovation along with green chemistry is a means of building a fast lane to achieve a competitive edge. Making social innovation the center of the business helps in creating the potential that will make a difference. Social innovation helps to preserve a healthy corporate culture. More and more organizations are opting to involve their workers in social innovation, often as a non-compulsory initiative or volunteer workforce. Executives have found it to be an important element in human resource management, and therefore it is becoming increasingly common. However, the greatest importance is measured by voluntary staff engagement and executive support. The integration of your team is one of the biggest benefits of involving employees in social innovations through employee volunteering. The main goal of the integration of employees is to merge the team. Usually, it is made up of people with completely different temperaments and character traits, which can lead to conflicts and tensions between them. However, if employees are given the opportunity to cooperate on something unrelated to work, it may transpire that disputes are fading away and a healthier atmosphere exists. Green chemistry will help the organizations to tap the new market that sustainability has opened and capture the targeted customers while making profits and at the same side saving the environment.

6.8 CONCLUSION

Green chemistry is the term used by many organizations nowadays, whenever an organization is following CSR then their only contribution to the society is monetary. Social innovation is known as a process that develops as well as deploys effective solutions for the challenging social and ecological issues that play a pivotal role in supporting social progress. Corporate social innovation engages itself in society-driven R&D and builds a true partnership between companies' agencies and NGOs, while corporate social responsibility is providing social and profitable services to the needy. Corporate social innovation with their partners tends to create something up-to-date and something inventive to provide a more sustainable solution for the people who need them. Social innovation is known as a key to market success.

It is hard to build an accountable corporate image. Treating social innovation as a natural way of operating in an enterprise enables organizations to establish and maintain a corporate image. Attempts are made to address social challenges using socially responsible solutions.

Social innovation follows a triple bottom-line (TBL) approach, which is a framework that recommends that companies commit themselves to social and environmental concerns just as much as they do to profits. The triple bottom line argues that instead of one bottom line, there should be three: profit, people, and the planet. A TBL seeks to gauge the level of commitment of a corporation to corporate social responsibility and its impact on the environment over time. This chapter also explores the social innovation's involvement in business and various key terms related to corporate social responsibility and corporate social innovation (a phenomenon that lacks conceptual knowledge).

In green chemistry exploration, developments have facilitated sustainable practices over time with contribution in ecologically appropriate logical and strategy practices in line with world needs for decades. In spite of these attempts, businesses should envision the financial feasibility of employing green chemistry in their procedures, which stops us from leveraging the use of this ideology. Investments and dissemination on the importance of green chemistry and how they affect directly, employees and patient health until environmental sustainability, are extremely important for the process of future improvements. Implementing green chemistry will require the joint efforts of all the employees in the organization at all levels as well as support from the consumers and the government.

REFERENCES

Carroll, S. J. (1984). Woman candidates and support for feminist concerns: The closet feminist syndrome. *Western Political Quarterly*, 37(2), 307–323.
Fabien Martinez, P. O. (2020). Perspectives on the role of business in social innovation. *Journal of Management Development*, 36(5), 681–695.
Gabriela, L. (2012). Social innovations in the context of modernization. *Sociológia*, 44, 291–313.
Ghosh, S. (2014). A study of the participation of the private sector companies of India in corporate social responsibility activities through conjoint analysis. *Global Business Review*, 18(2), 151–181.
Marco, B. A., Rechelo, B. S., Kogawa, E. G., & Salgado, H. R. (2018). Evolution of green chemistry and its multidimensional impacts: A review. *Saudi Pharmaceutical Journal*.
Paunescu, C. (2014). Current trends in social innovation research: Social capital, corporate social responsibility, impact measurement. *9*.
Rubalcaba, R. P. (2016). Social innovation research: An emerging area of innovation studies? *45*(9).
Saluja, Rajni Kapoor, Sangam. (2017). Corporate social responsibility—Evolution. *International Research Journal of Management Sociology & Humanities*, 8(11), 158–167.
Sheldon, R. A. (2017). Metrics of green chemistry and sustainability: Past, present and future. *ACS Sustainable Chemistry and Engineering*, 32–48.
Simon Slavin, T. M. (1993). *Community organization and social administration*. Routledge.
Stoian, C. (2016). Corporate social responsibility that "pays": A strategic approach to CSR for SME. *Journal of Small Business Management*, 55.
Tilt, C. A. (2016). Corporate social responsibility research: The importance of context. *Tilt International Journal of Corporate Social Responsibility*, 1.
Verma, P., & Singh, A. (2016). Fostering stakeholders trust through CSR reporting. *IIM Kozhikode Society and Management Review*, 5(2), 186–199.

7 Cellulose-Derived Levoglucosenone

Production and Use as a Renewable Synthon for the Production of Biobased Chiral Building Blocks, Solvents, Polymers, and Materials

Sami Fadlallah and Florent Allais

CONTENTS

DOI: 10.1201/9781003301769-7

7.1 INTRODUCTION

The transition from depleted fossil fuels to biomass resources is becoming more than ever a hot and urgent topic. Industries are under constant pressure to adopt economic and environmentally sustainable strategies in their manufacturing processes. For example, Unilever, one of the world's leading suppliers of food, cleaning, and personal care products, recently announced its "Clean Future" strategy to eliminate all non-renewable carbon from its products by 2030.[1] Similarly, Michelin, the French tire giant, announced its goal to make tires fully sustainable by 2050.[2] The continuous and growing demand for the use of biobased products is mainly driven by: (i) international and European political concerns and practical measures for the implementation of sustainable development,[3] (ii) growing consumer awareness on the importance of the use of biobased products,[4] and (iii) and the inevitable depletion of fossil resources accompanied by serious environmental problems that damage our planet and living facilities of future generations.[5] Nevertheless, a transition to a low-carbon economy and products requires the availability and reasonable price of raw materials, especially when it comes to industrial applications. The use of renewable, abundant, and cheap plant resources, such as lignocellulosic plants (e.g., wood) that contain a high proportion of cellulose, is a pioneering approach.[6] Although these renewable plant resources are abundant in nature, the extraction of the desired chemical platform products requires biorefineries to perform the conversion and separation of the different components of the biomass.[7]

Biomass pyrolysis is commonly used in biorefineries as one of the most promising technologies to convert biomass into solid products (charcoal), liquid bio-oil (that also contains hundreds of organic products)[8,9] and gaseous products.[10] This technique is receiving increasing attention from several studies, particularly due to its universality, renewability, and ease of transport.[11] Nonetheless, the reaction scheme of biomass pyrolysis is extremely complex due to the formation of many intermediates that complicate the separation process and increase the production cost of the products.[12] For this reason, selective pyrolysis was developed to control the process through the selection of suitable feedstocks, addition of a convenient catalyst, pretreatment of the feedstocks, and the selection of optimal conditions to carry out the biomass pyrolysis.[13,14]

In this context, levoglucosenone (LGO) (Figure 7.1) is an interesting chiral building block[15-17] produced from the acid-catalyzed flash pyrolysis of cellulosic waste such as bagasse and sawdust.[18-21] This simple and highly functional α,β-unsaturated ketone is an important key intermediate that can be used as a

FIGURE 7.1   Levoglucosenone (LGO) structure.

platform molecule for the synthesis of chemicals, solvents, polymers, and bio-based materials. This chapter describes the pyrolysis methods used to obtain levo-glucosenone from cellulosic biomass. In addition, since the use of solvent(s) is inevitable in most chemical reactions and since there is a need to find non-toxic and bio-renewable solvents,[22–24] this chapter also highlights the use of LGO as a feedstock to produce green solvents for diverse chemical processing. Finally, this chapter focuses on the rapidly evolving field of LGO-derived chiral synthons, as well as monomers and their utility in the synthesis of different classes of polymers via different polymerization methods.

7.2 PRODUCTION OF LEVOGLUCOSENONE

7.2.1 SYNTHETIC APPROACH FROM CELLULOSE

Accessing LGO from carbohydrates requires good control of the water content in the medium to prevent its subsequent dehydration into HMF (Scheme 7.1). Two main strategies can be implemented to keep LGO away from water: using a dehydrating agent or a polar solvent. Building on the latter, the dehydration of cellulose into LGO has been carried out in the presence of sulfuric acid and two polar aprotic solvents (i.e., γ-valerolactone and THF) and resulted in yields up to 95% from 1%w cellulose in THF at 210°C with 7.5 mM sulfuric acid.[25] The two strategies were combined and resulted in the use of dioxane and p-toluenesulfonic acid or methanesulfonic acid in the presence of 1%w phosphorous pentoxide (drying agent) (LGO yield up to 38% by treating 1%w cellulose in dioxane with 15 mM acid).[26]

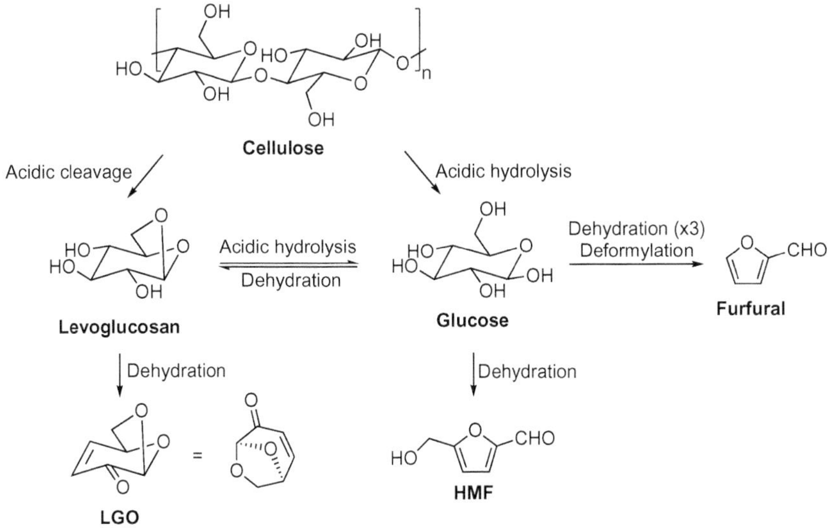

SCHEME 7.1 Production of LGO, levoglucosan, furfural and HMF from the acidic degradation of cellulose/glucose.

Pyrolyzing cellulose in the presence of metal catalysts has been widely studied. For instance, the pyrolysis of cellulose has been reported in the presence of several catalysts such as Ni and P co-modified mesoporous silicate-41 catalyst,[27] Mobil™ Composition of Matter (MCM)-41 and Fe-, Al-, and Cu-substituted samples in a fixed-bed glass reactor,[28] nanopowder titanium oxide, aluminum oxide, and aluminum titanate in a fixed-bed reactor,[29] and magnetic solid acid $Fe_3O_4/C-SO_3H_{600}$,[30] and provided LGO in 21.37%w, 53%w, 22%, and 20% yields, respectively.

The use of ionic liquid (IL) that act as a catalyst while avoiding the use of organic solvents is a very interesting strategy to prepare LGO from cellulose. The catalytic activity of various ionic liquids has been explored in the pyrolysis of cellulose for the production of LGO.[31,32] It was shown that not only sulfonate-based ILs provided the best LGO yields (up to 29.7%w), but they are also thermostable and can be readily recycled for subsequent runs.

Finally, non-conventional technologies such as microwave-[33] and plasma-activated pyrolysis[34] of cellulose have also been reported in the literature with LGO yields of 12.3w% and 43%, respectively.

7.2.2 Direct Biomass-Degradation Approach

Although there have been many LGO production processes relying on cellulose as starting material, producing LGO directly from lignocellulosic biomass is the most straightforward and economically relevant strategy as there is no need for any costly and energy- and/or chemical-consuming pretreatment of the lignocellulose to isolate cellulose.

The pyrolysis of a lignin-rich waste from softwood hydrolysis (a crude biorefinery waste stream) under microwave heating resulted in the conversion of the saccharides into LGO in an 8%w yield with high selectivity and to obtain purer lignin stream for subsequent valorizations.[21,35]

An extruder-based continuous fast catalytic pyrolysis of sawdust, from the pulp and paper industry for instance, in the presence of phosphoric acid and tetramethylene sulfone, was also reported in the literature.[36] To date, this technology is the only one that has been successfully scaled at the demonstration scale worldwide, currently producing LGO at the 50 T/year scale in Tasmania, and in a couple of years at the 100 T/year scale in the Grand Est region (France).

7.2.3 Synthetic Approach from Monosaccharide and Derivatives, as well as Non-sugar Compounds

The partial dehydration of the glucose units of lignocellulosic biomass is the main approach to access LGO, whereas a few reports on chemical routes starting from monosaccharides as well as a non-sugar compound can be found.

Levoglucosan has been the starting material of two chemical pathways. The first one consisted of its loading on activated carbon followed by the pyrolysis of the

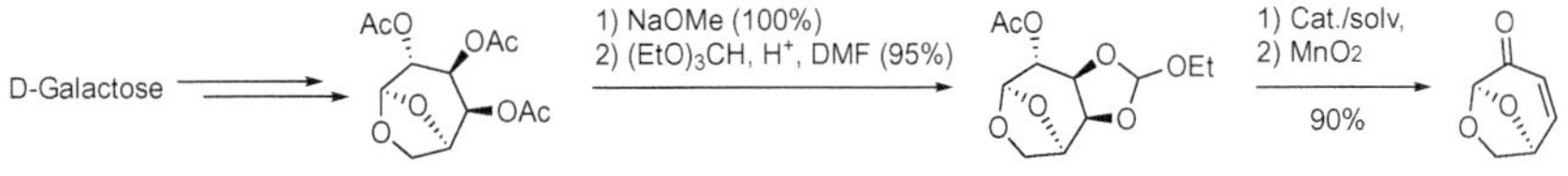

SCHEME 7.2   Synthesis of LGO from D-galactose.

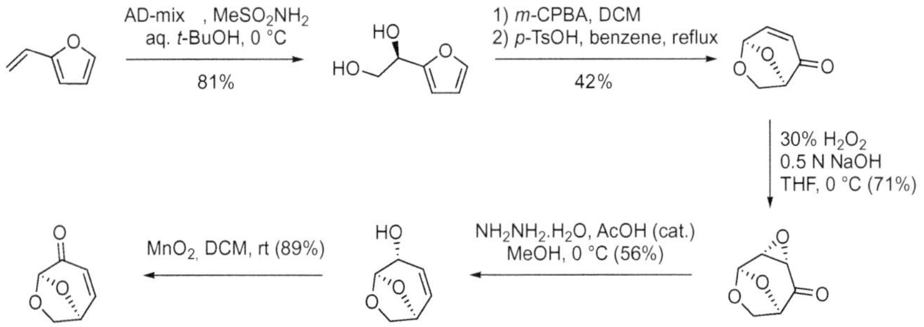

SCHEME 7.3   Enantioselective synthesis of LGO from 2-vinylfuran.

resulting material to achieve a 49% yield and high purity.[37] The second pathway, based on the use of propylsulfonic acid-functionalized silica catalysts in tetrahydrofuran, provided LGO up to 59% yield with complete levoglucosan conversion.[20]

An elegant route relying on the dehydration of glucose into a mixture of HMF and LGO in the presence of used chromium-tanned leather—acting as heterogeneous catalyst—was also reported in the literature.[38] This study demonstrated that classical metal catalysts may be advantageously replaced by this quite uncommon catalyst to provide LGO up to 12.8% yield at low chromium content.

A high-yielding eight-step pathway to LGO from D-galactose using a highly selective calcined zirconium oxide-catalyzed reductive decarboxylation as a key step has been also reported (ca. 48% overall yield) (Scheme 7.2).[39,40]

Finally, while the carbohydrate-based LGO production pathways benefit from the chiral pool of carbohydrates, an enantioselective multi-step route was reported from achiral 2-vinylfuran using the well-known stereoselective Sharpless dihydroxylation (39% over six steps, 93% e.e.) (Scheme 7.3).[41]

7.3   LEVOGLUCOSENONE-BASED BUILDING BLOCKS AND SOLVENTS

7.3.1   CYRENE™ AND CYGNETS™

LGO has been investigated for more than 50 years now as a valuable building block for the production of fine chemicals,[16,42] and more recently, the green solvent Cyrene™ for which two main synthetic procedures have been reported (Scheme 7.4). The first one is a classical palladium-catalyzed hydrogenation of the α,β-unsaturation of LGO developed by James Clark's group at the University of York (UK) that provides Cyrene™ in quantitative yield.[43] To offer a metal-free

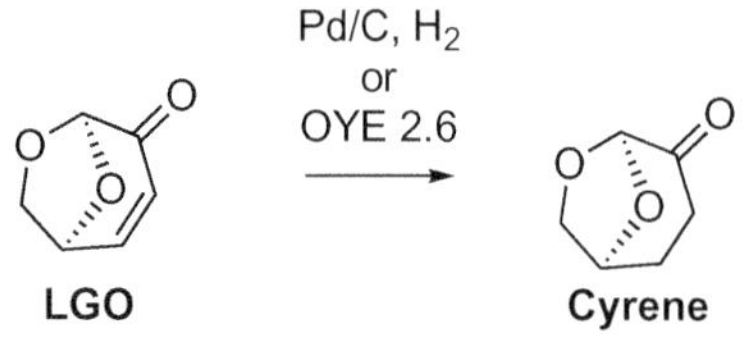

Pd/C, H₂

or

OYE 2.6

LGO → Cyrene

SCHEME 7.4 Chemical and biotechnological routes toward Cyrene™.

Cyrene™ needed for specific niche applications (e.g., electronics, drugs synthesis), our team—in collaboration with Prof. Jon Stewart's team at the University
of Florida (USA)—has designed and patented a biotechnological route involving
an ene-reductase (Old Yellow Enzyme 2.6) in a phosphate buffer. This biocatalytic reaction also proceeds in a quantitative yield.[44,45] It is noteworthy to mention
that Cyrene™ is REACH-compliant and thus recognized as a safe solvent. Since
its first report in the literature, several published works have demonstrated the
great potential of Cyrene™ as a sustainable alternative to toxic polar solvents,
e.g., NMP, in many applications,[46] such as liquid/liquid and liquid/solid extrac-
tions,[47,48] multilayer graphene production,[49] metal-organic frameworks (MOF)
synthesis,[50] amidification,[51,52] enzymatic reduction of α-ketoesters,[53] biomimetic
radical coupling,[54] polymerization,[55,56] or membrane synthesis.[57]

Building on the valuable properties of Cyrene™, Clark's group also investigated the synthesis of a new class of Cyrene™-derived solvents named Cygnet™
(Figure 7.2).[58] Readily obtained from the condensation of diols with the ketone
moiety, Cygnets™ are ketals of Cyrene™ that are in the same region as dichloromethane in the Hansen space, making them potential substitutes for toxic chlorinated solvents. To differentiate the different Cygnets™, a nomenclature has been
established according to the number of carbons of the substituents at positions
4 and 5 of the dioxolane ring (e.g., Cygnet™ deriving from ethylene glycol and
n-1,2-hexanediol are Cygnet™ 0.0 and Cygnet™ 4.0, respectively). Cygnet™ 0.0
has been recently used in combination with Cyrene™ as a safer alternative to
chlorinated solvents for the preparation of membranes from cellulose acetate,
polysulfone, and polyimide, and it was shown that, by playing with the Cygnet™/
Cyrene™ ratio, one can tune the properties of the resulting membranes. In the
same paper, the authors also demonstrated that Cygnet™–Cyrene™ blends were

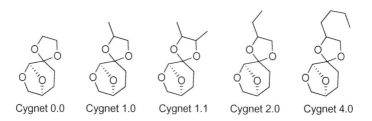

Cygnet 0.0   Cygnet 1.0   Cygnet 1.1   Cygnet 2.0   Cygnet 4.0

FIGURE 7.2 Cygnet solvents.

suitable for substituting diphenyl ether in the biocatalyzed synthesis of high molecular weight aliphatic and furanic polyesters.[57]

7.3.2   LGO-Derived Chiral Building Blocks

Besides Cyrene™ and Cygnets™, a great deal of work has been dedicated to the development of LGO-derived high-value chiral building blocks, for instance, 4(S)-γ-hydroxymethyl-α,β-butenolide   (HBO)   and   (S)-γ-hydroxymethyl-γ-butyrolactone (2H-HBO), which are valuable chiral lactones.[59] HBO and 2H-HBO can be readily obtained from LGO through a two-step pathway involving Baeyer–Villiger oxidation (BVO) and acidic hydrolysis, and a three-step pathway involving BVO/hydrolysis and palladium-catalyzed hydrogenation (or vice versa), respectively (Scheme 7.5). Although high yielding, the first two synthetic routes reported in the literature were not entirely sustainable. Indeed, the very first synthesis of HBO and H-HBO by Koseki et al. consisted of a peracid-mediated BVO (explosive) in dichloromethane followed by hydrolysis in acidic methanol (and an extra palladium-catalyzed hydrogenation for 2H-HBO).[60] The second route reported by Corma et al. replaced peracids, DCM, and methanol with tin-zeolites, 1,4-dioxane, and Amberlyst-15, respectively.[61] Although greener, this pathway still uses a rather hazardous solvent and catalyst. Convinced of the great potential of HBO and 2H-HBO, our group devised and patented no less than three sustainable processes to offer green access to these two chiral lactones from LGO and/or Cyrene™: (i) a lipase-mediated Baeyer–Villiger oxidation (BVO) followed by acidic hydrolysis in ethanol,[62] (ii) an organic solvent- and catalyst-free BVO that does not require any hydrolysis,[63] and (iii) a cyclohexene-monooxygenase (CHMO)-mediated BVO of Cyrene™.[64] It is worth mentioning that the organic solvent- and catalyst-free BVO has been successfully scaled up at the kilo-scale.

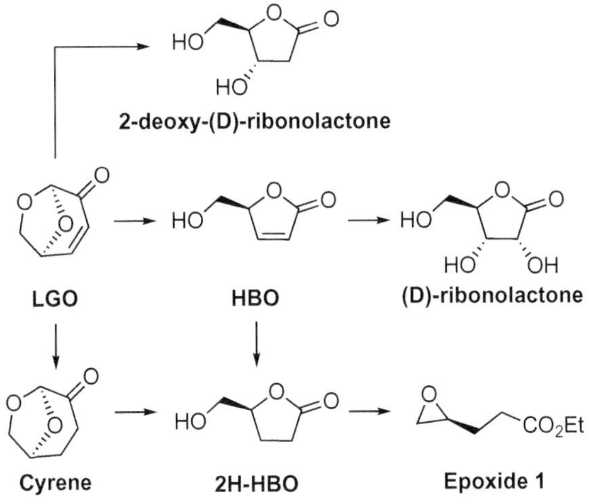

SCHEME 7.5   LGO-derived high-value chiral building blocks and Cyrene™.

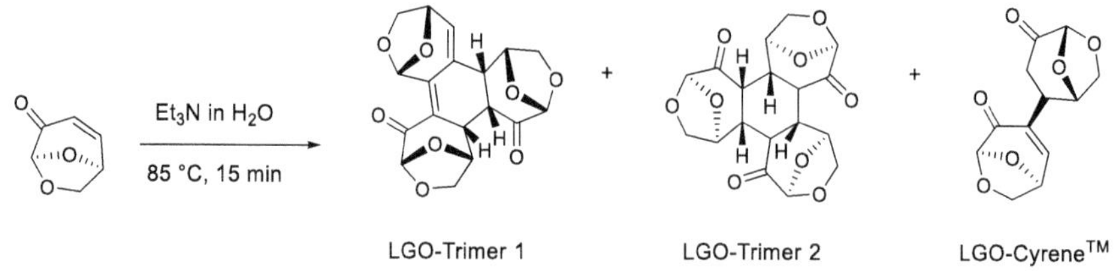

LGO-Trimer 1   LGO-Trimer 2   LGO-Cyrene™

SCHEME 7.6   Base-catalyzed oligomerization of LGO.

From HBO, three other valuable building blocks, chiral epoxide 1,[65] (D)-ribono-lactone,[66] and 2-deoxy-(D)-ribonalactone[67] were also successfully synthesized (Scheme 7.5).

7.4   LEVOGLUCOSENONE-BASED MONOMERS AND POLYMERS

LGO has recently been used as a versatile chiral platform for the synthesis of partially and fully renewable monomers and polymers.[22] Direct polymerization of LGO without any modification has not been successful so far. Indeed, Shafizadeh et al. showed that base-catalyzed oligomerization of LGO is possible by heating it in aqueous triethylamine to give only two trimers (olefinic and non-olefinic) and a dimeric product as shown in Scheme 7.6.[68] Possible mechanisms for the condensation of three LGO units to form LGO-Trimer 1 and LGO-Trimer 2 by Michael additions are reported in the original work.

Organic chemistry can be used as a powerful tool to efficiently modify LGO structure and access easily polymerizable monomers. In the following section—which is divided according to the type of LGO polymers into polyacrylates, polyacetals, polycyclic olefins, polyesters, polycarbonates, and poly(vinylether lactone)—the synthesis of LGO-derived monomers and corresponding polymers is presented.

7.4.1   POLYACRYLATES

Polyacrylates are widely used in a large number of industrial processes to produce adhesives, paints, coatings, films, as well as in medical applications such as bone cement and dentistry (e.g., artificial teeth, jaws, and a variety of fillings).[69,70] The use of these materials in various everyday products is due to their thermal stability,[71] attractive mechanical properties, high glass transition temperature (T_g), and transparency.[72] The strong growth in the polyacrylates market is mainly led by materials produced from monomers derived from fossil fuels that are often volatile and toxic.[73] In addition, the unsustainable methods adopted to produce acrylate monomers and the corresponding polymers contribute greatly to the generation of huge amounts of waste and increased greenhouse gas emissions.[23,24] Therefore, there is a need to find renewable and non-toxic monomer building blocks that can be polymerized through environmentally friendly and efficient routes to produce eco-friendly polyacrylates.[74,75] In this context, LGO plays an important role as a highly versatile biobased platform molecule to produce monomers and polymers from biomass.[22,74] The first two publications describing the synthesis of polymers from LGO aimed at synthesizing polyacrylates (Scheme 7.7).[76,77]

The alkene bond of the BVO product of LGO, i.e., HBO, was reduced by dihydrogen (H_2) in the presence of a catalytic amount of 10% palladium on carbon (Pd/C) to give 2H-HBO in 87% yield (Scheme 7.7).[62] The latter was used by Saito et al. as a precursor to produce the first LGO-derived acrylate monomer, m-2H-HBO.[76] More precisely, m-2H-HBO was synthesized in 79% yield by reacting 2H-HBO with methacrylic anhydride (MAN) in ethyl acetate in the presence of 4-dimethylaminopyridine (DMAP) and triethyl amine (TEA) (Scheme 7.7).[76] Later that year, Allais et al. reported a greener chemoenzymatic synthesis of m-2H-HBO using *Candida antarctica* type B lipase (CAL-B, aka Novozyme 435) to access m-2H-HBO in 62% yield.[77]

The ability of m-2H-HBO to be polymerized by free-radical polymerization was first investigated by Saito et al.[76] They showed that only 46% of the corresponding polyacrylate, poly(m-2H-HBO) (Scheme 7.7), could be isolated if no solvent was used to conduct the polymerization. In addition, polymers with high dispersity were isolated (Đ up to 4.2), which shows the uncontrolled behavior of

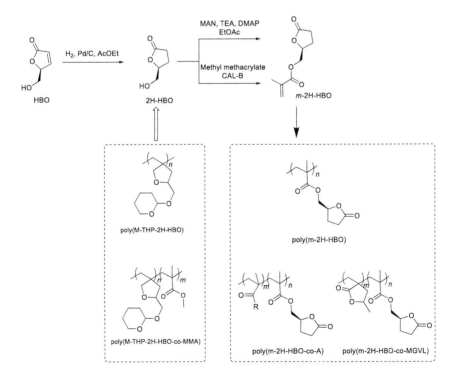

SCHEME 7.7 Synthesis of the first HBO-derived acrylate monomer and its use in polymerization.

polymer growth during polymerization.[76] Although solvent-free polymerization is generally considered a more environmentally friendly route to produce polymers with low waste generation, the aforementioned low conversion/yield as well as the poor control of the polymerization show the importance of using a suitable solvent. For this reason, the same authors tried radical solution polymerizations of m-2H-HBO in different solvents such as xylene, toluene, ethyl acetate, n-butanol, ethylene glycol, and dimethylsulfoxide. Regardless of the difference in toxicity of the different solvents, the best yield (78%) was obtained in the case of dimethylsulfoxide (DMSO), which led to polymers with a number average molecular weight (M_n) of 28.6 kDa and $Đ$ = 3.47.[76]

On the other hand, Allais et al. also recorded the free-radical polymerization of m-2H-HBO (synthesized by the chemoenzymatic route) under conditions similar to those reported by Saito et al. (1 mol% AIBN, 1 M monomer in toluene, 75°C, 90 min).[77] Nevertheless, better results were obtained due to the improved yield of the polymer described as quantitative in the original publication. In addition, Allais et al. tested the activity of other radical initiators such as V-65 and V-70 for the polymerization of m-2H-HBO; however, lower yields were obtained (43% and 37%, respectively).[77]

The thermal behavior of LGO-derived renewable polyacrylates was found to be very similar to that of Poly(methacrylic acidmethyl ester (PMMA), whose annual production exceeded 2.8 million tons in 2011,[70] and thus 2H-HBO can be considered a suitable candidate to replace methyl methacrylate (MMA).[78] Specifically, the polymers produced possessed T_g values ~96°C and thermal stability up to ~200°C with 20% degradation ($T_{20\%}$) at 300°C. Furthermore, very interestingly, the copolymerization of m-2H-HBO with methacrylamide (MAA) (R = NH$_2$) was carried out in dimethylformamide and led to high T_g (up to 187°C) when 89% of the MAA was incorporated in the backbone chain.[76] In addition, m-2H-HBO was successfully copolymerized with the biobased monomer methylene-γ-valerolactone (MGVL), also yielding valuable copolymers with comparable high T_g values.[77]

In the same work, Allais et al. performed the protection of the hydroxy group of 2H-HBO with 3,4-dihydro-2H-pyran to give 2-tetrahydropyranyl (THP)-protected THP-2H-HBO. Methylenation of the newly formed molecules was achieved followed by free-radical polymerization to give homopolymers and copolymers (with methyl methacrylate) with the same T_g ~ 148°C (Scheme 7.7). The deprotection of the corresponding homopolymer took place at 240°C after 90 min and resulted in a functional polymer containing hydroxy groups that can undergo post-polymerization modification.

As previously mentioned, the hydrogenation of the conjugated double bond of LGO gives access to Cyrene™.[44,62] Saito et al. employed a synthetic two-step procedure to produce another interesting acrylate monomer using Cyrene™ as a precursor (Scheme 7.8).[79] The first step involved a reduction of the carbonyl moiety that led to a mixture of *endo-* and *exo-*isomers of levoglucosenol (HO-Cyrene™) isolated in 95% yield and a 70% enantiomeric excess in favor of the *exo-*isomer. These isomers were not separated and were used as a mixture for

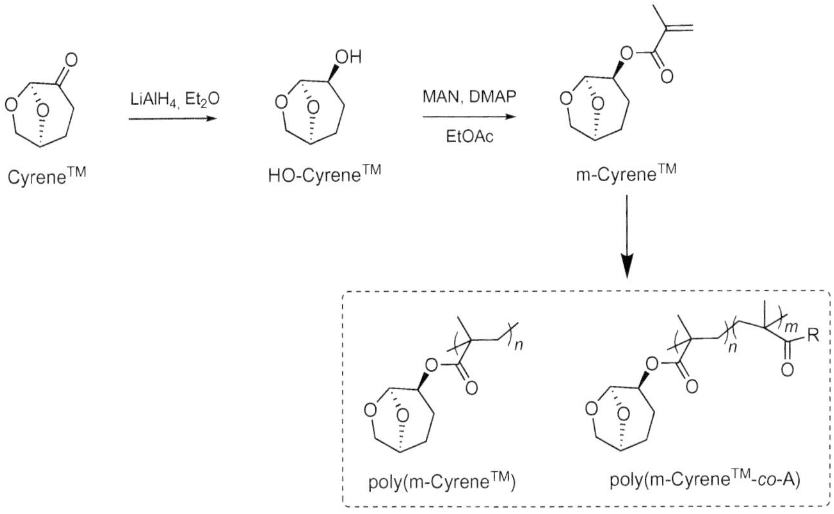

SCHEME 7.8   Synthesis of Cyrene™ polyacrylates monomer.

the transesterification reaction with MAN to give the target acrylate monomer (m-Cyrene™) in 84% yield.[79]

The tendency of m-Cyrene™ to undergo free-radical polymerization in the presence of AIBN was introduced by Saito et al. and successfully led to high-value polymers, poly(m-Cyrene™), while keeping the bicyclic acetal rings intact. The authors tried different reaction conditions and studied their impact on polymer formation. Once again, bulk polymerization led to the lowest polymer yield 45%. Solution polymerizations of m-Cyrene™ were carried out in different solvent γ-gamma valerolactone and Cyrene™. In all cases, better yields and polymers with higher molecular weights were obtained especially when Cyrene™ was used as reaction solvent (85% and 38 kDa). Furthermore, copolymerization reactions of m-Cyrene™ with common methacrylic monomers such as isobornyl methacrylate were also performed in Cyrene™ solution and proved successful.[79] The thermal properties of the Cyrene™-derived acrylate polymers were studied and showed high thermal stability (20% degradation at 290°C) and high T_g up to ~210°C. Such a range makes these polymers the best reported vinyl polymers to date, especially for their commercial use in industrial applications.[79]

7.4.2   POLYACETALS

Polyacetal or polyoxymethylene is a tough, solvent- and fuel-resistant engineering thermoplastic widely used in producing high precision parts, which require high lubricity (low friction) and rigidity, as well as excellent dimensional stability.[80] In recent years, biobased and biodegradable polyacetals from biomass have gained renewed interest due to their high oxygen-to-carbon ratio (carbon bonded to two −OR groups). Indeed, biomass is an oxygen-rich resource that can potentially supply many carbonyl and polyol monomers for polyacetals formation.[80] In this context, Schlaad et al. prepared fully renewable thermoplastic polyacetals from LGO.[81–84] Slight modifications of the LGO structure were required to prepare polylevoglucosenone by ring-opening metathesis polymerization (ROMP)[81,84] or (photo-)cationic ring-opening polymerization[82,83] (Scheme 7.9). The same research group also attempted to produce polymers from LGO via radical or anionic polymerization, but at best only oligomers were isolated.

Early attempts by Debsharma, Schlaad et al. to polymerize the conjugated double bond of LGO by ROMP failed due to the existence of a polar functional group, i.e., carbonyl that hinders the reactivity of the metathesis catalyst.[81] Indeed, it is known that the tolerance of metathesis catalysts to polar functional groups greatly affects their reactivity toward the ROMP of strained olefins.[85] Thus, a protection step of the functional group(s) adjacent to the cyclic olefin is necessary prior to ROMP to avoid catalyst deactivation.[85] To make LGO polymerizable by ROMP, Schlaad et al. reduced the keto portion of LGO using $NaBH_4$ in water as a solvent to quantitatively obtain HO-LGO. This was followed by the protection of the hydroxy group to prepare a series of biomass-derived levoglucosenyl alkyl ethers (alkyl = methyl, ethyl, n-propyl, isopropyl, and n-butyl) (Scheme 7.9).[81,84]

The authors screened six different Ru-based metathesis catalysts, initially for the ROMP of HO-LGO, where only a derivative of the second-generation Grubbs catalyst bearing a mono-*ortho*-substituted N-heterocarbon (NHC) as a ligand (Grubbs catalyst second generation (GII), Scheme 7.9) was found to be

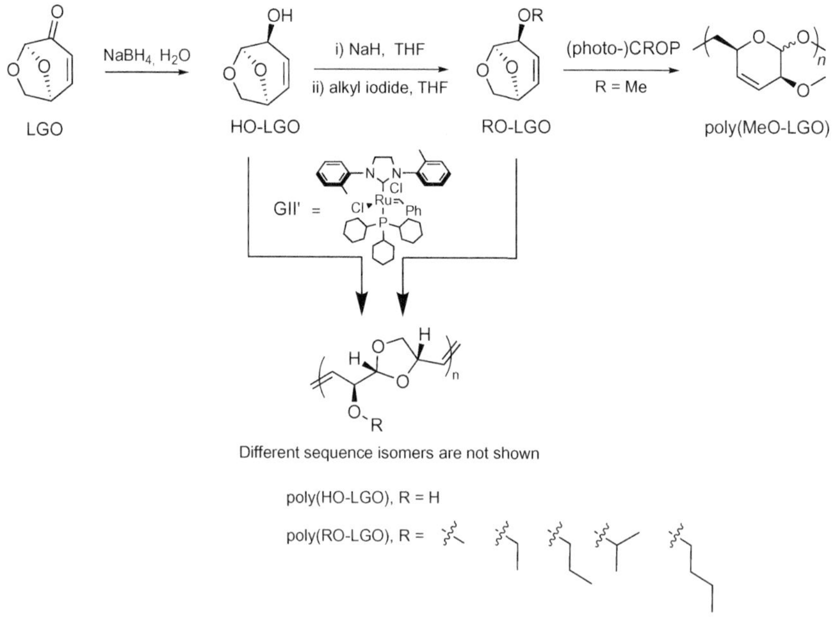

SCHEME 7.9 Synthesis of thermoplastic polyacetals from LGO.

effective. The authors explained this unexpected reactivity by the lower steric hindrance of the mono-ortho NHC ligand around the Ru center. In the first paper of this work,[81] 1,4-dioxane was used to perform ROMP at room temperature. However, 1,4-dioxane is a toxic solvent, so they later reported the use of "green" aprotic polar solvents such as 2-methyltetrahydrofuran, dihydrolevoglucosenone (Cyrene™) and ethyl acetate.[84] In addition, they prepared a series of levoglucosenyl alkyl ethers as shown in Scheme 7.9 and studied their effect on the polymer properties.[84]

Different poly(RO-LGO) isomers were observed due to the chirality of the constituent monomers (RO-LGO) that can be incorporated in different orientations to the growing polymer chain, thus leading to the formation of head-to-head, head-to-tail, and tail-to-tail configurations.[81] An apparent degree of polymerization of ~100 and dispersity of ~2 were observed for all polyacetals produced, which we also found to be amorphous with thermoplastic behavior.[84] Furthermore, more interestingly, the choice of the alkyl substituent (R) showed a direct effect on the T_g of the produced polymer. Specifically, the T_g decreased from 100°C to 0°C with an increase in the length of the alkyl substituent of the polymers.[84]

In addition, the presence of functional groups such as hydroxy (in the case of poly(HO-LGO)) and olefin groups makes the polymer susceptible to modification after polymerization. For example, hydrogenation of the double bonds of poly(HO-LGO) was achieved in 98% yield using H_2/Pd-C.[81] These polymers also contain degradable acetal moieties that completely degrade in 40 days at room temperature in a 1,4-dioxane solution containing acidic water.[81]

On the other hand, Debsharma et al. reported the cationic ring-opening polymerization (CROP) of LGO and showed that protection of HO-LGO is necessary for the polymerization to proceed via Lewis acid-catalyzed CROP.[82] Thus, MeO-LGO was prepared by deprotonation of the hydroxy using sodium hydride, followed by methylation with methyl iodide, and isolated in high yield (81%) (Scheme 7.9). Two initiators were then tested for the CROP of MeO-LGO: triflic acid (TfOH) and boron trifluoride etherate (BF_3.OEt$_2$). TfOH was found to be more efficient than BF_3.OEt$_2$ for CROP in dichloromethane at room temperature and 0°C, resulting in polymers of high molecular weight and yield. A full explanation is presented in the original publication, including a mechanism for the proposed proton-initiated CROP pathway. Unlike the polyacetal polymers obtained by ROMP,[81,84] the CROP polymer (annotated as poly(MeO-LGO) in Scheme 7.9) contained only one sequence isomer.[82]

Regarding the thermal analysis of poly(MeO-LGO), the results showed the formation of semi-crystalline thermoplastic polymers with T_g ~ 35°C.[82] Interestingly, this behavior is different from the amorphous polyacetals of poly(HO-LGO) obtained by ROMP.[81] This shows the high impact of the polymerization method on the structures and properties of the formed polymer which, in turn, determines the corresponding target applications.

The same authors also investigated the direct and sensitized photocationic ring-opening polymerization (photo-CROP) of levoglucosenyl methyl ether (MeO-LGO).[83] They showed that polyacetals can be produced from the latter through photochemistry under mild and low energy conditions. UV and visible photolysis of diphenyliodonium hexafluorophosphate (DPI) provided acids or carbocations capable of inducing CROP to lead into high molecular weight polymers when polymerizations were conducted at −15°C. They also showed that the polymerization exhibited a living character and can be used to prepare block copolymers of MeO-LGO with cyclohexene oxide and isobutyl vinyl ether by sequential monomer addition. The mechanisms of the DPI-mediated CROP under UV and visible light irradiation are reported in the original work.[83]

7.4.3 Polycyclic Olefins

Polycyclic olefin, or cyclic olefin polymer, is a relatively new class of polymers compared to commodities such as polypropylene and polyethylene. It can be prepared by vinyl addition polymerization or ROMP of strained cyclic olefins, such as dicyclopentadiene or norbornene.[86] Cyclic olefin polymers and their copolymers are very important industrial materials that can be used in a wide variety of applications, including packaging, healthcare, optics, electronics, and medical devices.[87–90]

   Due to the highly reactive double bond of LGO, norbornene-containing LGO molecules can be prepared by the Diels–Alder reaction of LGO with cyclopentadiene (Cp). Shafizadeh et al.[91] and Horton et al.[92] reported the synthesis of this molecule in the 1980s and showed that LGO can act as a dienophile to give two isomers, a major *endo*-N-LGO (called N-LGO in this chapter) and a minor *exo*-N-LGO. Banwell et al. reported the synthesis of a library of monomers derived from N-LGO and N-isoLGO (isoLGO being the pseudo-enantiomer of N-LGO) (Schemes 7.10 and 7.11).[93] The synthesis of these precursors (N-LGO and N-isoLGO) was performed in dichloromethane in 70% yield. In the same work,[93] new hydroxyl-containing derivatives of N-LGO were also prepared by

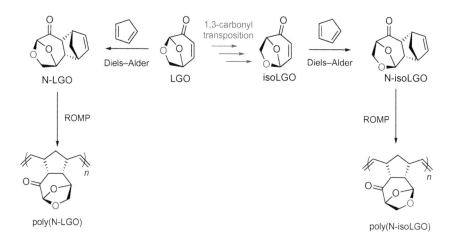

SCHEME 7.10   Synthesis of norbornene-containing LGO monomers and polymers.

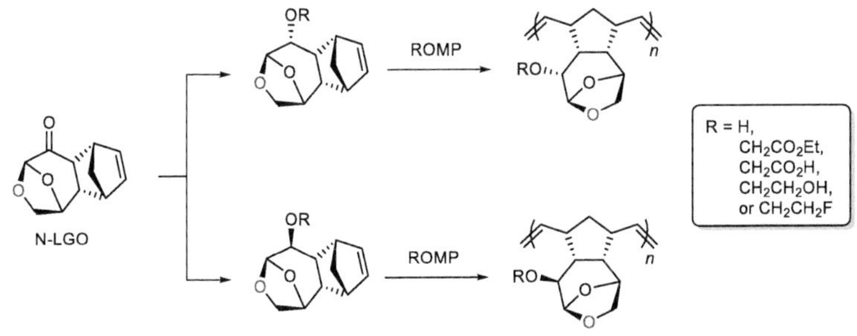

SCHEME 7.11 Synthesis of norbornene-containing epimeric derivatives from N-LGO and the corresponding ROMP polymers.

the reduction of the corresponding ketone of N-LGO using $NaBH_4$ in an aqueous ethanol solution in 92% yield. The resulting two epimeric alcohols were isolated and reacted with ethyl α-bromoacetate in dry tetrahydrofuran to produce the ester derivatives. These were also reduced to the corresponding carboxylic acids using lithium hydroxide in methanol or to the corresponding alcohols using lithium borohydride in tetrahydrofuran. Finally, fluorinated analogs were prepared through fluorination using diethylaminosulfur trifluoride in dichlormethane.[93]

The performance of the norbornene isoLGO molecule in ROMP reactions was first studied by Banwell et al. using the following metathesis catalysts: second-generation Grubbs, third-generation Grubbs, and second-generation Hoveyda-Grubbs and Stewart-Grubbs.[93] Among these four catalysts, GIII showed the best control over the polymerization process. Thus, GIII was chosen to study the ROMP of N-LGO, N-isoLGO, and their derivatives.

As these polymers have the potential to be implemented on an industrial scale, concentrated ROMPs of N-LGO using the cheapest first-generation Grubbs catalyst were performed by Allais et al.[94] This study paved the way for more environmentally friendly reactions by decreasing the environmental factor (E factor) of the polymerization procedure from 6.0 to 2.8 kg waste/kg poly(N-LGO). In addition, the same authors investigated the effect of the stereochemistry of the N-LGO isomers (i.e., *endo*-N-LGO and *exo*-N-LGO) on the polymerization and showed that ROMP proceeds faster with the *exo*-isomer.[94]

TGA analysis of the prepared polymers showed that they are thermostable up to 360°C and 168°C in the case of the N-LGO[94] and N-isoLGO[93] based polymers, respectively. Notably, the ability of pseudo-enantiomeric monomers (i.e., N-LGO and iso-N-LGO) to form thermostable copolymers with a $T_{d5\%}$ up to 278°C was also proven.[93]

On the other hand, previous studies showed that the H_3PO_4-catalyzed and H_2O_2-mediated BVO of N-LGO leads into the hydroxy-lactone (N-HBO).[92,94,95] A polar solvent such as isopropanol was required to increase the rate of the oxidation reaction.[96] Allais et al. studied the ROMP of N-HBO without protecting the hydroxy group of HBO to obtain poly(N-HBO) with $T_{d5\%}$ up to 380°C (Scheme 7.12).[94] Indeed, although avoiding protection of the hydroxyl group led

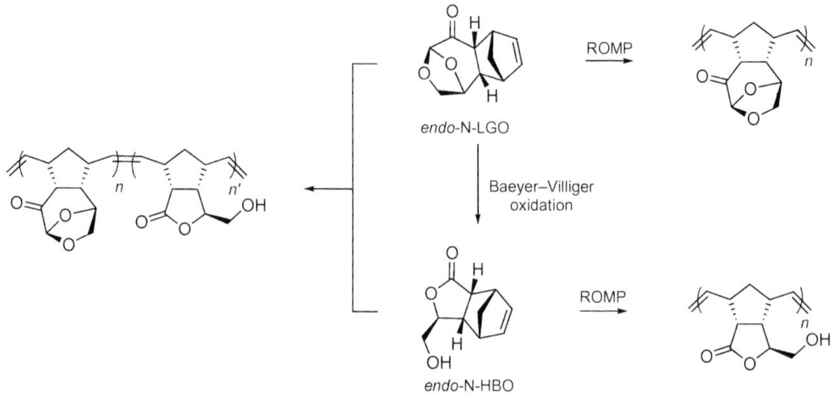

SCHEME 7.12   Synthesis of diverse functional polymers from norbornene LGO monomers.

to some complications during polymerization, such an approach decreases the waste generated by avoiding protection and deprotection steps, but also provides direct access to free hydroxy-based functional polymers from simply prepared monomers such as N-HBO. The same authors also successfully prepared copolymers containing both LGO and HBO moieties by the random copolymerization of N-LGO and N-HBO under mild conditions.[94]

Allais et al. extended the family of norbornene LGO molecules to include a bifunctional methacrylate monomer that can be readily synthesized through a chemoenzymatic pathway. The enzyme-catalyzed transesterification of the hydroxyl group of N-HBO with vinyl methacrylate (3 equiv.) was performed in acetone at 50°C using an immobilized *Candida antarctica* Lipase B (Novozyme® 435)[97] to provide norbornene-(S)-γ-hydroxymethyl-α,β-butenolide methacrylate (N-HBO-MA) in 90% yield (Scheme 7.13).[55] The authors showed that, in the presence of a metathesis catalyst, the methacrylate group remained intact and only the norbornene moiety was polymerized to yield functional poly(N-HBO-MA).

Furthermore, Cyrene™ was successfully introduced as a biobased solvent for ROMP reactions using highly active catalysts such as second-generation Hoveyda-Grubbs catalyst.[55] Also, Allais et al. showed that the production poly(N-HBO-MA) in Cyrene™ is possible at g-scale. The potential of Cyrene™ as a green

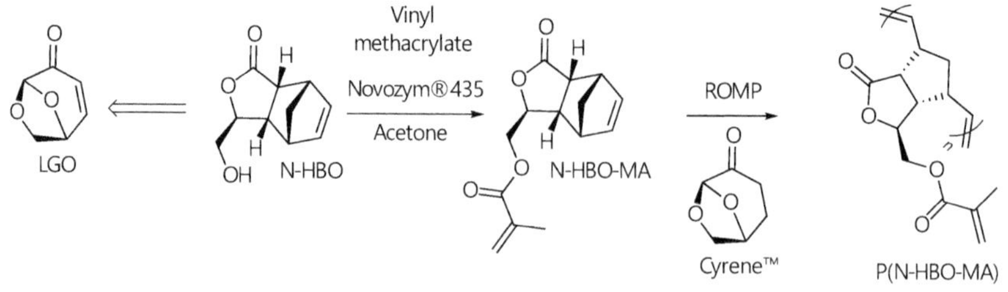

SCHEME 7.13   Synthesis of N-HBO-MA and its ROMP in Cyrene™.

solvent for ROMP of N-LGO and N-HBO was investigated as well and found to be successful.[55] Such exceptional tolerance of Cyrene™ toward functional monomers warrants further attention to this compound as a "general" green solvent for a broader range of polymer synthesis. The thermal analysis of the prepared poly(N-HBO-MA) showed high thermal stability with $T_{d5\%}$ up to 401°C which is the highest known for LGO-derived polymers. Furthermore, dynamic mechanical analysis of films prepared from poly(N-HBO-MA) by solvent casting showed a low T_g of −16.8°C.[55]

7.4.4 POLYESTERS

Polyesters are an extremely important class of polymers that not only meet the market requirements in terms of desirable qualities but also in production cost, making it one of the most economically important commodity polymers.[98] For example, the annual global production of polyethylene terephthalate (PET), which is widely used as a thermoplastic polymer exceeded 70 Mt in 2018 with annual growth of more than 4%.[99] One route for the production of polyesters is the polycondensation of diesters (or diacids) and diols (or polyols). This method is characterized by the wide availability of raw materials, making it a versatile method for a wide range of polyesters.

In this context, Allais et al. designed several diol and triol monomers from LGO.[67,100–102] The first diol derived from LGO is shown in Scheme 7.14. Its synthesis started with the optimization of the dimerization reaction of LGO, which was initially described by Shafizadeh et al. in 1982.[68] Specifically, Allais et al. showed that by dissolving LGO in methanol in the presence of a catalytic amount of K_2CO_3, a dimer LGO-Cyrene™ can be formed in quantitative yield, *via* the Michael addition of LGO onto itself.[100] A mechanism for its formation is described in the original work.[100] The catalyst- and solvent-free BVO of LGO-Cyrene™ afforded a bicyclic diol named 2H-HBO-HBO in 60% yield. The solvent-free polycondensation of this bicyclic diol with diacyl chlorides was then studied to produce the first LGO-derived polyesters.[100] Different diacyl chlorides were employed, ranging from malonyl chloride (m = 1) to adipoyl chloride (m = 4), in addition to the aromatic monomer terephthaloyl chloride, to produce polymers with distinct thermal properties. It is important to note that polymers with

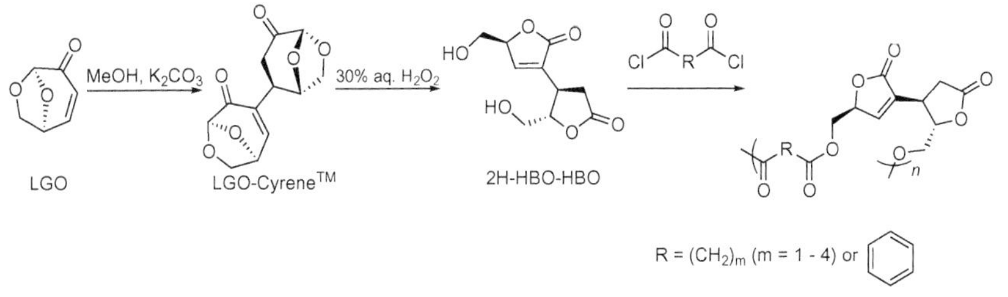

R = (CH₂)ₘ (m = 1 - 4) or [benzene ring]

SCHEME 7.14   Synthesis of the first LGO-derived polyesters.

low yields ranging from 38% to 45% were obtained. Indeed, although solvent-free polymerization is more environmentally friendly as it limits the use of solvents and in turn the amount of waste generated from a synthetic procedure, it is usually accompanied by drawbacks, especially in the case of polycondensation including the low control of the process and the occurrence of side reactions that decrease the yield and molecular weights of the materials produced.[23] Thermal analyses of these LGO-derived polymers showed high stability with $T_{d50\%}$ up to 300°C. Moreover, as expected, T_g was influenced by the aliphatic chain length of the diacyl chlorides (e.g., T_g reached 71°C when succinyl chloride was engaged and 81°C in the case of terephthaloyl chloride).[100]

Another important reaction that LGO can undergo is the hydration by Michael addition of water onto its conjugated double bond under acidic and basic conditions (Scheme 7.15).[103] Although this synthesis allowed facile access to HO-HBO, it required a high amount of base (i.e., Et$_3$N). Thus, Allais et al. developed a more sustainable synthesis of HO-LGO by replacing Et$_3$N with K$_3$PO$_4$ (using 0.05 equiv.) and performing the reaction in water.[67] These conditions allowed higher yields (up to 82%) and also limited the dimerization of LGO, a competitive side reaction that can occur under basic conditions and leads to the previously described LGO-Cyrene™.[67] On the other hand, the synthesis of hydrated HBO molecule (2-deoxy-D-ribonolactone, HO-HBO) was described in 1995 in high yield (85%) but using a large excess of peracetic acid and dimethyl sulfide.[104] Allais et al. revisited this reaction and performed the green one-pot catalyst- and organic solvent-free BVO of HO-LGO to access HO-HBO in 79% yield.[67] Afterwards, they assessed the potential of this diol for the production of polyesters. The solvent-free polycondensation of HO-HBO with aliphatic diacyl chlorides led to oligomers having molecular weights in the range of 2.3–4.5 kDa with T_g values between −21 and −2°C and melting temperatures (T_m) from 87°C to 144°C.[100]

Allais et al. then became interested in developing another fully renewable 5-membered cyclic molecule, HBO-citro, which was prepared through a one-pot, two-step synthesis from LGO: oxa-Michael addition of citronellol followed by BVO of the crude product (Scheme 7.16).[101,102] The reduction of the lactone moiety of HBO-citro with 2 equiv. of NaBH$_4$ in tetrahydrofuran at room temperature led to the formation of a triol derivative bearing citronellol as a side group (Triol-citro). The authors tried to replace tetrahydrofuran (a non-renewable and toxic solvent) with a green solvent such as 2-MeTHF (a sugar-derived solvent). They also performed the selective reduction of HBO-citro to synthesize a diol derivative using diisobutylaluminum hydride (DIBAL-H) as a reducing agent.

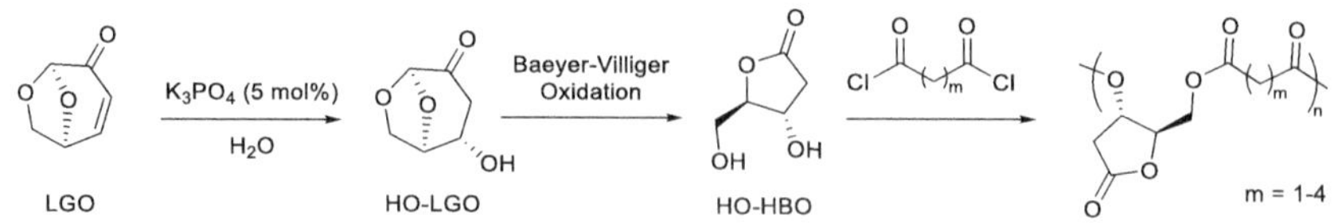

SCHEME 7.15   Synthesis of HO-HBO-based polyesters from LGO.

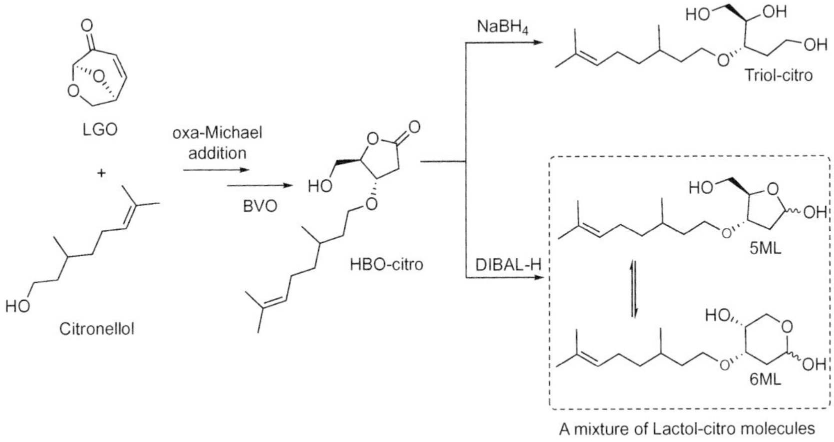

SCHEME 7.16 Synthesis of Triol-citro and Lactols-citro from LGO.

Nevertheless, NMR analysis showed the presence of a mixture of 5- and 6-membered cyclic forms of Lactol-citro (Scheme 7.16). The isolation of one pure product of the lactols was not possible.[101,102]

Functional renewable (co)polyesters with citronellol pendant chains and low T_g (as low as −67°C) were then prepared from Triol-citro and Lactols-citro (5-membered and 6-membered) via polycondensation reactions with diacyl chlorides having different chain lengths ($m = 1–4$) (Scheme 7.17).[101,102] Having T_g values in this negative range is very attractive for applications that require a T_g below body temperature such as in the biomedical field. Furthermore, Allais et al. studied the

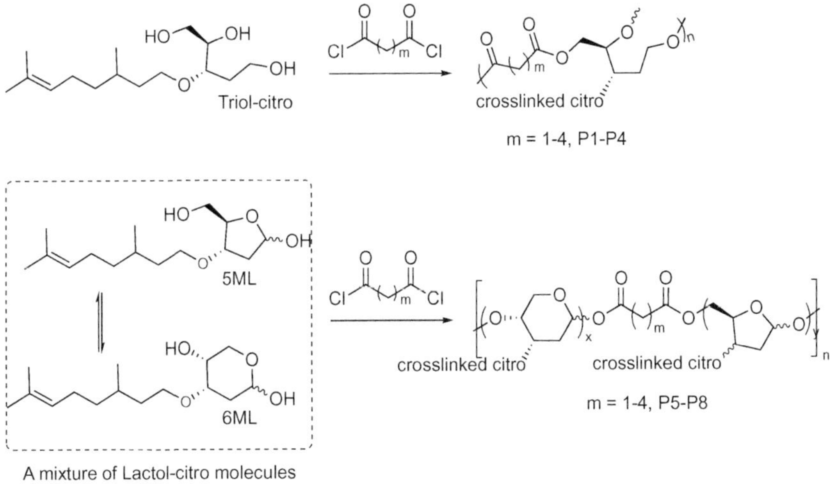

SCHEME 7.17 Synthesis of LGO-derived polyesters containing citronellol pendant group.

enzymatic degradation of P1–P8 (Scheme 7.17) using a commercial lipase from *Thermomyces lanuginosus* (Lipopan® 50 BG; Novozymes®, Danmark). When P1–P4 were exposed to Lipopan® 50 BG, an 80% degradation of P2–P4 was observed after 80 h.[101] On the other hand, when P5–P8 was subjected to enzymatic degradation under the same condition, a higher degradation rate was found compared to those obtained by the polycondensation reactions of diacyl chlorides with Triol-citro (P1–P4).[102] The authors explained this behavior by the possibility that a highly crosslinked/branched structure, as in the case of P1–P4, can induce a globular polymeric conformation that hinders the accessibility of the enzyme to the inner surface of the chains and delays/prevents the hydrolysis of the ester bonds.[102]

7.4.5 POLYCARBONATES

Polycarbonates are engineering thermoplastics[105] that are widely used in diverse applications such as electrical instruments,[106] safety helmets, optical plates, and many others.[107–110] The annual global demand for polycarbonates exceeds 4.4 million metric tons and is increasingly growing.[111] An attractive approach to produce polycarbonates is through the polycondensation of hydroxy-containing monomers (diols or polyols) and organic carbonates such as dimethyl carbonate.[112] This method is endowed by the large availability of the biobased hydroxyl and carbonate substrates. Allais et al. took advantage of the presence of three hydroxyls in Triol-citro and prepared functional photocrosslinkable polycarbonates by the solvent-free metal-catalyzed polycondensation with isosorbide-derived dimethylcarbonate (DCI) (Scheme 7.18).[113] It is important to note that isosorbide is a

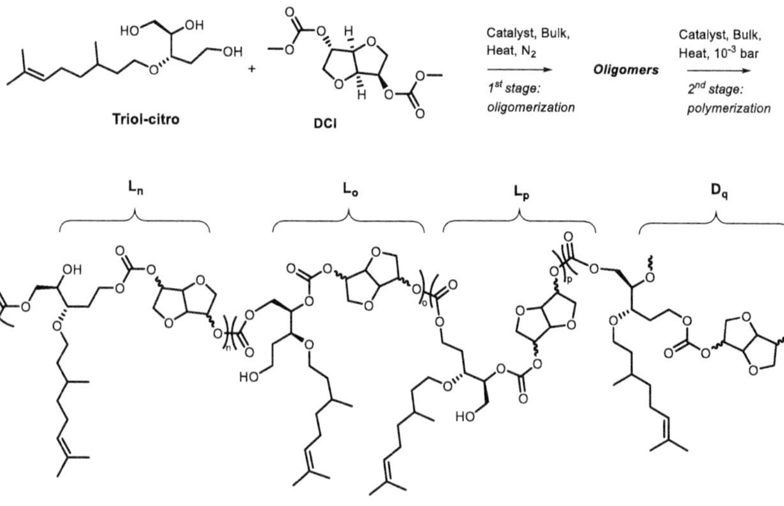

SCHEME 7.18 Synthesis of photocrosslinkable polycarbonates through one-pot bulk polycondensation of Triol-citro and DCI.

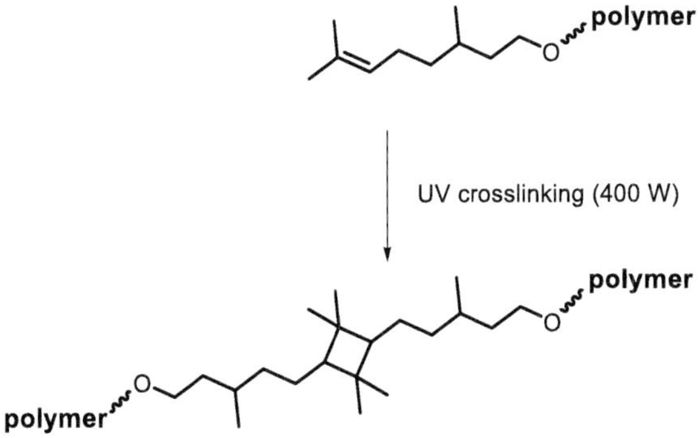

SCHEME 7.19 Representation of the UV-crosslinked citronellol moiety.

biobased platform chemical[114–117] that has been employed in the synthesis of poly-carbonates,[118,119] as well as polyesters[120] and polyurethanes.[121]

As shown in Scheme 7.18, Triol-citro is a chiral molecule that has three hydroxy groups with different reactivities. In addition, DCI contains two *endo*- and *exo*-carbonate groups that react differently. So, the polycondensation of Triol-citro and DCI leads to a very complex polymer structure. The microstructure was elucidated by Allais et al. and showed the existence of four repeating units: three linear units with pendant hydroxy groups (L_n, L_o, and L_p) and one branched unit (D_q) resulting from the reaction of the three hydroxy groups of Triol-citro with the two carbonates of DCI.[113] The effect of different conditions, including catalyst type, temperature, and monomer concentration, on the percentage of the four different units was also tested. These polycarbonates exhibited low glass transition temperatures (T_g as low as −72°C) and $T_{d50\%}$ up to 159°C.

On the other hand, since these polymers contain citronellol side chains, they studied the UV crosslinking of the terminal double bonds of citronellol (Scheme 7.19). The results showed that a longer UV exposure time led to a higher cross-linking rate. Moreover, the thermal analysis showed the formation of a stiffer/less flexible polymer structure (an increase of T_g from −60°C to −6°C, and $T_{d50\%}$ from 212°C to 444°C), thanks to crosslinking.[113]

7.4.6 POLY(VINYLETHER LACTONE)

Following the work on the preparation of polyesters via solvent-free polycon-densation of 2H-HBO-HBO with diacyl chlorides, Allais et al. investigated the synthesis of polycarbonates by reacting the 2H-HBO-HBO monomer with di(methylcarbonate) isosorbide (DCI).[122] All metal-based polycondensation tri-als failed to allow the condensation of the monomers together and to achieve the wanted polycarbonates. Indeed, the polymer formed but they found no con-densation reaction between 2H-HBO-HBO hydroxy groups and DCI carbonate moieties.[122]

The authors tried to elucidate the mechanism and found that the homopolymerization of 2H-HBO-HBO occurred in the presence of a basic catalyst at 120°C and led to a 100% renewable hyperbranched polymer. Being rather fast, this formation prevented any polycondensation between the hydroxy groups of 2H-HBO-HBO and DCI when the condensation polymerization was sought. Driven by these results, they also tried to homopolymerize HBO (without any modification) and obtained functional polymer materials with E factor as less as 0.03 kg of waste/kg of poly(HBO).

For more insights on the polymerization mechanism, the same group attempted the homopolymerization of 2H-HBO (hydrogenated derivative of HBO) and HBO-Bn (benzyl-protected derivative of HBO) under similar conditions. The trials showed that the endocyclic double bond of HBO is crucial for the polymerization to occur and a free hydroxy group is necessary for the propagation to proceed. By careful interpretation of NMR spectra, they concluded that hyperbranched/crosslinked materials of poly(2H-HBO-HBO) and poly(HBO) were obtained through Michael addition, oxa-Michael addition, and proton transfer processes to lead to a very complex poly(vinylether lactone) complex copolymeric structures (Scheme 7.20).

Poly(2H-HBO-HBO) showed a relatively high T_g (i.e., 82°C), which was found to decrease as the molecular weight of the polymer decreased. Furthermore, a 5% thermal degradation of the hyperbranched polymers appeared in the range of 222–243°C. The thermal analysis of poly(HBO) showed that T_g existed in the

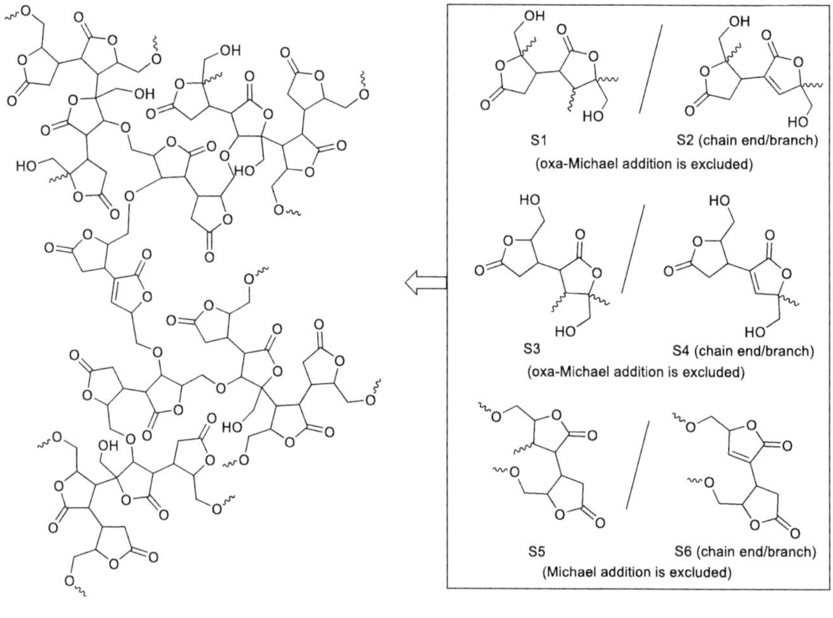

SCHEME 7.20   Possible hyperbranched structure of poly(vinylether lactone) obtained from the homopolymerization of 2H-HBO-HBO.

range of 45–86°C depending on the catalyst presence and the molecular weight of the polymer as well. However, lower thermal stability, when compared to poly(2H-HBO-HBO), was observed with $T_{d5\%}$ of poly(HBO) in the range of 141–170°C.[122]

7.5 CONCLUSION

LGO offers an enormous opportunity to be effectively used in the fine chemical and commodity industries. The number of methods dedicated to the production and valorization of this cellulose-derived chiral molecule is constantly increasing. In addition, the range of applications is expanding to include its use in the production of synthons, solvents, monomers, and polymers, as well as bioactive molecules. The high potential of LGO lies not only in its ability to be chemically modified thanks to the presence of readily reactive functional moieties, but also in the ease of its preparation by flash pyrolysis of cellulosic waste. Pyrolysis is a low-cost technology that can be easily scaled up and applied to a wide variety of feedstocks such as cellulose. In addition, this method does not require the use of solvents, which significantly reduces the amount of waste generated by the manufacturing processes. Nevertheless, solvents are crucial substances that in most cases are unavoidable to improve the chemical reactivity and carry out chemical syntheses, so the efficient design and production (in terms of quantity and cost) of biobased, non-toxic, and environmentally safer organic solvents—such as Cyrene™—is of extreme importance. Cyrene™ is commonly used today as a REACH-compliant green aprotic polar solvent for various synthetic reactions. In addition, Cyrene™ was used as a starting molecule to produce Cygnet 0.0 as a safer replacement for polar aprotic solvents suitable for conducting enzymatic polycondensation reactions for polymer synthesis. Finally, the potential of LGO as a biobased key intermediate to produce biobased chiral synthons, monomers, and polymers was highlighted. Different polymerization techniques such as free-radical, ring-opening, and condensation polymerizations were employed to produce different types of LGO-derived polymers including polyacrylates, polyacetals, polycyclic olefins, polyesters, polycarbonates, and poly(vinylether lactones).

REFERENCES

1. *Clean Home. Clean Planet. Clean Future.* Unilever. https://www.unilever.com/brands/home-care/clean-future/ (accessed 2022-07-11).
2. *Michelin - In 2050, Michelin Tires Will Be 100% Sustainable.* Michelin. https://www.michelin.com/en/press-releases/in-2050-michelin-tires-will-be-100-sustainable/ (accessed 2022-07-11).
3. El Bilali, H.; Strassner, C.; Ben Hassen, T. Sustainable Agri-Food Systems: Environment, Economy, Society, and Policy. *Sustainability* 2021, *13*(11), 6260. https://doi.org/10.3390/su13116260.
4. Gaffey, J.; McMahon, H.; Marsh, E.; Vehmas, K.; Kymäläinen, T.; Vos, J. Understanding Consumer Perspectives of Bio-based Products—A Comparative Case Study from Ireland and the Netherlands. *Sustainability* 2021, *13*(11), 6062. https://doi.org/10.3390/su13116062.

5. Höök, M.; Tang, X. Depletion of Fossil Fuels and Anthropogenic Climate Change—A Review. *Energy Policy* 2013, *52*, 797–809. https://doi.org/10.1016/j.enpol.2012.10.046.
6. Narayan, R. Biomass (Renewable) Resources for Production of Materials, Chemicals, and Fuels. In: *Emerging Technologies for Materials and Chemicals from Biomass; ACS Symposium Series.* American Chemical Society, 1992, *476*, 1–10. https://doi.org/10.1021/bk-1992-0476.ch001.
7. Gallezot, P. Conversion of Biomass to Selected Chemical Products. *Chem. Soc. Rev.* 2012, *41*(4), 1538–1558. https://doi.org/10.1039/C1CS15147A.
8. Wang, H.; Male, J.; Wang, Y. Recent Advances in Hydrotreating of Pyrolysis Bio-oil and Its Oxygen-Containing Model Compounds. *ACS Catal.* 2013, *3*(5), 1047–1070. https://doi.org/10.1021/cs400069z.
9. Adhikari, S.; Srinivasan, V.; Fasina, O. Catalytic Pyrolysis of Raw and Thermally Treated Lignin Using Different Acidic Zeolites. *Energy Fuels* 2014, *28*(7), 4532–4538. https://doi.org/10.1021/ef500902x.
10. Liu, C.; Wang, H.; Karim, A. M.; Sun, J.; Wang, Y. Catalytic Fast Pyrolysis of Lignocellulosic Biomass. *Chem. Soc. Rev.* 2014, *43*(22), 7594–7623. https://doi.org/10.1039/C3CS60414D.
11. Guo, M.; Song, W.; Buhain, J. Bioenergy and Biofuels: History, Status, and Perspective. *Renew. Sustain. Energy Rev.* 2015, *42*, 712–725. https://doi.org/10.1016/j.rser.2014.10.013.
12. Babu, B. V. Biomass Pyrolysis: A State-of-the-Art Review. *Biofuels Bioprod. Biorefin.* 2008, *2*(5), 393–414. https://doi.org/10.1002/bbb.92.
13. Di Blasi, C.; Branca, C.; Galgano, A.; D'Agostino, P. Thermal Behavior of Beech Wood during Sulfuric Acid Catalyzed Pyrolysis. *Energy Fuels* 2015, *29*(10), 6476–6484. https://doi.org/10.1021/acs.energyfuels.5b01315.
14. Kan, T.; Strezov, V.; Evans, T. J. Lignocellulosic Biomass Pyrolysis: A Review of Product Properties and Effects of Pyrolysis Parameters. *Renew. Sustain. Energy Rev.* 2016, *57*, 1126–1140. https://doi.org/10.1016/j.rser.2015.12.185.
15. Ariel, M. S.; Maria, M. Z.; Rolando, A. S. Recent Applications of Levoglucosenone as Chiral Synthon. *Curr. Org. Synth.* 2012, *9*(4), 439–459.
16. Comba, M. B.; Tsai, Y.; Sarotti, A. M.; Mangione, M. I.; Suárez, A. G.; Spanevello, R. A. Levoglucosenone and Its New Applications: Valorization of Cellulose Residues. *Eur. J. Org. Chem.* 2018, *5*(5), 590–604. https://doi.org/10.1002/ejoc.201701227.
17. Kühlborn, J.; Groß, J.; Opatz, T. Making Natural Products from Renewable Feedstocks: Back to the Roots? *Nat. Prod. Rep.* 2020, *37*(3), 380–424. https://doi.org/10.1039/C9NP00040B.
18. Halpern, Y.; Riffer, R.; Broido, A. Levoglucosenone (1,6-Anhydro-3,4-Dideoxy-.DELTA.3-.Beta.-D-Pyranosen-2-One). Major Product of the Acid-Catalyzed Pyrolysis of Cellulose and Related Carbohydrates. *J. Org. Chem.* 1973, *38*(2), 204–209. https://doi.org/10.1021/jo00942a005.
19. Sui, X.; Wang, Z.; Liao, B.; Zhang, Y.; Guo, Q. Preparation of Levoglucosenone through Sulfuric Acid Promoted Pyrolysis of Bagasse at Low Temperature. *Bioresour. Technol.* 2012, *103*(1), 466–469. https://doi.org/10.1016/j.biortech.2011.10.010.
20. Oyola-Rivera, O.; He, J.; Huber, G. W.; Dumesic, J. A.; Cardona-Martínez, N. Catalytic Dehydration of Levoglucosan to Levoglucosenone Using Brønsted Solid Acid Catalysts in Tetrahydrofuran. *Green Chem.* 2019, *21*(18), 4988–4999. https://doi.org/10.1039/C9GC01526D.
21. Clark, J. H.; Bruyn, M. D.; Budarin, V. L. Method for Producing Levoglucosenone. WO2016170329A1, October 27, 2016.

22. Fadlallah, S.; Mouterde, L. M. M.; Garnier, G.; Saito, K.; Allais, F. Cellulose-Derived Levoglucosenone, a Great Versatile Chemical Platform for the Production of Renewable Monomers and Polymers. In: *ACS Symposium Series*; Cheng, H. N., Gross, R. A., Eds.; American Chemical Society, Washington, DC, 2020, *1373*, 77–97. https://doi.org/10.1021/bk-2020-1373.ch005.

23. Fadlallah, S.; Sinha Roy, P.; Garnier, G.; Saito, K.; Allais, F. Are Lignin-Derived Monomers and Polymers Truly Sustainable? An In-Depth Green Metrics Calculations Approach. *Green Chem.* 2021, *23*(4), 1495–1535. https://doi.org/10.1039/D0GC03982A.

24. El Itawi, H.; Fadlallah, S.; Allais, F.; Perré, P. Green Assessment of Polymer Microparticles Production Processes: A Critical Review. *Green Chem.* 2022, *24*(11), 4237–4269. https://doi.org/10.1039/D2GC00578F.

25. Cao, F.; Schwartz, T. J.; McClelland, D. J.; Krishna, S. H.; Dumesic, J. A.; Huber, G. W. Dehydration of Cellulose to Levoglucosenone Using Polar Aprotic Solvents. *Energy Environ. Sci.* 2015, *8*(6), 1808–1815. https://doi.org/10.1039/C5EE00353A.

26. Cao, Q.; Ye, T.; Li, W.; Chen, J.; Lu, Y.; Gan, H.; Wu, H.; Cao, F.; Wei, P.; Ouyang, P. Dehydration of Saccharides to Anhydro-Sugars in Dioxane: Effect of Reactants, Acidic Strength and Water Removal In Situ. *Cellulose* 2020, *27*(17), 9825–9838. https://doi.org/10.1007/s10570-020-03490-2.

27. Li, K.; Wang, B.; Bolatibieke, D.; Nan, D.-H.; Zhang, Z.-X.; Cui, M.-S.; Lu, Q. Catalytic Fast Pyrolysis of Biomass with Ni-P-MCM-41 to Selectively Produce Levoglucosenone. *J. Anal. Appl. Pyrol.* 2020, *148*, 104824. https://doi.org/10.1016/j.jaap.2020.104824.

28. Casoni, A. I.; Nievas, M. L.; Moyano, E. L.; Álvarez, M.; Diez, A.; Dennehy, M.; Volpe, M. A. Catalytic Pyrolysis of Cellulose Using MCM-41 Type Catalysts. *Appl. Catal. Gen.* 2016, *514*, 235–240. https://doi.org/10.1016/j.apcata.2016.01.017.

29. Fabbri, D.; Torri, C.; Mancini, I. Pyrolysis of Cellulose Catalysed by Nanopowder Metal Oxides: Production and Characterisation of a Chiral Hydroxylactone and Its Role as Building Block. *Green Chem.* 2007, *9*(12), 1374–1379. https://doi.org/10.1039/B707943E.

30. Qian, L.; Xu, F.; Liu, S.; Lv, G.; Jiang, L.; Su, T.; Wang, Y.; Zhao, Z. Selective Production of Levoglucosenone by Catalytic Pyrolysis of Cellulose Mixed with Magnetic Solid Acid. *Cellulose* 2021, *28*(12), 7579–7592. https://doi.org/10.1007/s10570-021-04010-6.

31. Kudo, S.; Zhou, Z.; Norinaga, K.; Hayashi, J. Efficient Levoglucosenone Production by Catalytic Pyrolysis of Cellulose Mixed with Ionic Liquid. *Green Chem.* 2011, *13*(11), 3306–3311. https://doi.org/10.1039/C1GC15975E.

32. Kudo, S.; Zhou, Z.; Yamasaki, K.; Norinaga, K.; Hayashi, J. Sulfonate Ionic Liquid as a Stable and Active Catalyst for Levoglucosenone Production from Saccharides via Catalytic Pyrolysis. *Catalysts* 2013, *3*(4), 757–773. https://doi.org/10.3390/catal3040757.

33. Doroshenko, A.; Pylypenko, I.; Heaton, K.; Cowling, S.; Clark, J.; Budarin, V. Selective Microwave-Assisted Pyrolysis of Cellulose towards Levoglucosenone with Clay Catalysts. *ChemSusChem* 2019, *12*(24), 5224–5227. https://doi.org/10.1002/cssc.201903026.

34. Lusi, A.; Radhakrishnan, H.; Hu, H.; Hu, H.; Bai, X. Plasma Electrolysis of Cellulose in Polar Aprotic Solvents for Production of Levoglucosenone. *Green Chem.* 2020, *22*(22), 7871–7883. https://doi.org/10.1039/D0GC02813D.

35. Bruyn, M. D.; Fan, J.; Budarin, V. L.; Macquarrie, D. J.; Gomez, L. D.; Simister, R.; Farmer, T. J.; Raverty, W. D.; McQueen-Mason, S. J.; Clark, J. H. A New Perspective in Bio-refining: Levoglucosenone and Cleaner Lignin from Waste Biorefinery Hydrolysis Lignin by Selective Conversion of Residual Saccharides. *Energy Environ. Sci.* 2016, *9*(8), 2571–2574. https://doi.org/10.1039/C6EE01352J.

36. Court, G. R.; Lawrence, C. H.; Raverty, W. D.; Duncan, A. J. Method for Converting Lignocellulosic Materials into Useful Chemicals. US20120111714A1, May 10, 2012.

37. Yang, Z. et al. CN113372398, October 10, 2021.

38. Bobbink, F. D.; Huang, Z.; Menoud, F.; Dyson, P. J. Leather-Promoted Transformation of Glucose into 5-Hydroxymethylfurfural and Levoglucosenone. *ChemSusChem* 2019, *12*(7), 1437–1442. https://doi.org/10.1002/cssc.201802830.

39. Shibagaki, M.; Takahashi, K.; Kuno, H.; Honda, I.; Matsushita, H. Synthesis of Levoglucosenone. *Chem. Lett.* 1990, *19*(2), 307–310. https://doi.org/10.1246/cl.1990.307.

40. Shibagaki, M. C. J. T. I.; Takahashi, K. C. J. T. I.; Kuno, H. C. J. T. I.; Honda, I. C. J. T. I.; Mori, M. C. J. T. I.; Matsushita, H. C. J. T. I. Method of Preparing Levoglucosenone. EP0416577A2, March 13, 1991.

41. Taniguchi, T.; Nakamura, K.; Ogasawara, K. Non-carbohydrate Route to Levoglucosenone and Its Enantiomer Employing Asymmetric Dihydroxylation. *Synlett* 1996, *10*(10), 971–972. https://doi.org/10.1055/s-1996-5652.

42. Camp, J. E.; Greatrex, B. W. Levoglucosenone: Bio-based Platform for Drug Discovery. *Front. Chem.* 2022, *10*.

43. Sherwood, J.; Bruyn, M. D.; Constantinou, A.; Moity, L.; McElroy, C. R.; Farmer, T. J.; Duncan, T.; Raverty, W.; Hunt, A. J.; Clark, J. H. Dihydrolevoglucosenone (Cyrene) as a Bio-based Alternative for Dipolar Aprotic Solvents. *Chem. Commun. (Camb)* 2014, *50*(68), 9650–9652. https://doi.org/10.1039/C4CC04133J.

44. Mouterde, L. M. M.; Allais, F.; Stewart, J. D. Enzymatic Reduction of Levoglucosenone by an Alkene Reductase (OYE 2.6): A Sustainable Metal- and Dihydrogen-Free Access to the Bio-based Solvent Cyrene®. *Green Chem.* 2018, *20*(24), 5528–5532. https://doi.org/10.1039/C8GC03146K.

45. Stewart, J. D.; Allais, F.; Mouterde, L. M. M. Materials and Methods for Alkene Reduction of Levoglucosenone by an Alkene Reductase. WO2018183706A1, October 4, 2018.

46. Camp, J. E. Bio-Available Solvent Cyrene: Synthesis, Derivatization, and Applications. *ChemSusChem* 2018, *11*(18), 3048–3055. https://doi.org/10.1002/cssc.201801420.

47. Duval, A.; Avérous, L. Dihydrolevoglucosenone (Cyrene™) as a Versatile Biobased Solvent for Lignin Fractionation, Processing, and Chemistry. *Green Chem.* 2022, *24*(1), 338–349. https://doi.org/10.1039/D1GC03395F.

48. Brouwer, T.; Schuur, B. Dihydrolevoglucosenone (Cyrene), a Biobased Solvent for Liquid–Liquid Extraction Applications. *ACS Sustain. Chem. Eng.* 2020, *8*(39), 14807–14817. https://doi.org/10.1021/acssuschemeng.0c04159.

49. Pan, K.; Fan, Y.; Leng, T.; Li, J.; Xin, Z.; Zhang, J.; Hao, L.; Gallop, J.; Novoselov, K. S.; Hu, Z. Sustainable Production of Highly Conductive Multilayer Graphene Ink for Wireless Connectivity and IoT Applications. *Nat. Commun.* 2018, *9*(1), 5197. https://doi.org/10.1038/s41467-018-07632-w.

50. Zhang, J.; White, G. B.; Ryan, M. D.; Hunt, A. J.; Katz, M. J. Dihydrolevoglucosenone (Cyrene) as a Green Alternative to N,N-dimethylformamide (DMF) in MOF Synthesis. *ACS Sustain. Chem. Eng.* 2016, *4*(12), 7186–7192. https://doi.org/10.1021/acssuschemeng.6b02115.

51. Wilson, K. L.; Murray, J.; Jamieson, C.; Watson, A. J. B. Cyrene as a Bio-based Solvent for HATU Mediated Amide Coupling. *Org. Biomol. Chem.* 2018, *16*(16), 2851–2854. https://doi.org/10.1039/C8OB00653A.

52. Bousfield, T. W.; Pearce, K. P. R.; Nyamini, S. B.; Angelis-Dimakis, A.; Camp, J. E. Synthesis of Amides from Acid Chlorides and Amines in the Bio-based Solvent Cyrene™. *Green Chem.* 2019, *21*(13), 3675–3681. https://doi.org/10.1039/C9GC01180C.

53. de Gonzalo, G. Biocatalysed Reductions of α-Ketoesters Employing CyreneTM as Cosolvent. *Biocatal. Biotransform.* 2022, *40*(4), 252–257. https://doi.org/10.1080/10242422.2021.1887150.

54. Mention, M. M.; Flourat, A. L.; Peyrot, C.; Allais, F. Biomimetic Regioselective and High-Yielding Cu(I)-Catalyzed Dimerization of Sinapate Esters in Green Solvent Cyrene™: Towards Sustainable Antioxidant and Anti-UV Ingredients. *Green Chem.* 2020, *22*(6), 2077–2085. https://doi.org/10.1039/D0GC00122H.

55. Fadlallah, S.; Peru, A. A. M.; Longé, L.; Allais, F. Chemo-Enzymatic Synthesis of a Levoglucosenone-Derived Bi-functional Monomer and Its Ring-Opening Metathesis Polymerization in the Green Solvent Cyrene™. *Polym. Chem.* 2020, *11*(47), 7471–7475. https://doi.org/10.1039/D0PY01471K.

56. Grune, C.; Thamm, J.; Werz, O.; Fischer, D. Cyrene™ as an Alternative Sustainable Solvent for the Preparation of Poly(Lactic-Co-glycolic Acid) Nanoparticles. *J. Pharm. Sci.* 2021, *110*(2), 959–964. https://doi.org/10.1016/j.xphs.2020.11.031.

57. Milescu, R. A.; Zhenova, A.; Vastano, M.; Gammons, R.; Lin, S.; Lau, C. H.; Clark, J. H.; McElroy, C. R.; Pellis, A. Polymer Chemistry Applications of Cyrene and Its Derivative Cygnet 0.0 as Safer Replacements for Polar Aprotic Solvents. *ChemSusChem* 2021, *14*(16), 3367–3381. https://doi.org/10.1002/cssc.202101125.

58. Alves Costa Pacheco, A.; Sherwood, J.; Zhenova, A.; McElroy, C. R.; Hunt, A. J.; Parker, H. L.; Farmer, T. J.; Constantinou, A.; De bruyn, M.; Whitwood, A. C.; Raverty, W.; Clark, J. H. Intelligent Approach to Solvent Substitution: The Identification of a New Class of Levoglucosenone Derivatives. *ChemSusChem* 2016, *9*(24), 3503–3512. https://doi.org/10.1002/cssc.201600795.

59. Flourat, A. L.; Haudrechy, A.; Allais, F.; Renault, J.-H. (S)-γ-Hydroxymethyl-α,β-Butenolide, a Valuable Chiral Synthon: Syntheses, Reactivity, and Applications. *Org. Process Res. Dev.* 2019. https://doi.org/10.1021/acs.oprd.9b00468.

60. Koseki, K.; Ebata, T.; Kawakami, H.; Matsushita, H.; Itoh, K.; Naoi, Y. Method of Preparing (S)-γ-Hydroxymethyl-α,β-Butenolid. US49947585, February 19, 1991.

61. Paris, C.; Moliner, M.; Corma, A. Metal-Containing Zeolites as Efficient Catalysts for the Transformation of Highly Valuable Chiral Biomass-Derived Products. *Green Chem.* 2013, *15*(8), 2101–2109. https://doi.org/10.1039/C3GC40267C.

62. (a) Flourat, A. L.; Peru, A. A. M.; Teixeira, A. R. S.; Brunissen, F.; Allais, F. Chemo-Enzymatic Synthesis of Key Intermediates (S)-γ-Hydroxymethyl-α,β-Butenolide and (S)-γ-Hydroxymethyl-γ-Butyrolactone via Lipase-Mediated Baeyer–Villiger Oxidation of Levoglucosenone. *Green Chem.* 2014, *17*(1), 404–412. https://doi.org/10.1039/C4GC01231C. (b) Allais, F.; Flourat, A.; Peru, A.; Teixeira, A.; Brunissen, F.; Spinnler, H. E. Method for Transforming 4-Hydroxymethyl Butyrolactone or 4-Hydroxymethyl Butenolide US201515307712, June 1, 2017.

63. (a) Bonneau, G.; Peru, A. A. M.; Flourat, A. L.; Allais, F. Organic Solvent- and Catalyst-Free Baeyer–Villiger Oxidation of Levoglucosenone and Dihydrolevoglucosenone (Cyrene®): A Sustainable Route to (*S*)-γ-Hydroxymethyl-α,β-Butenolide and (*S*)-γ-Hydroxymethyl-γ-Butyrolactone. *Green Chem.* 2018, *20*(11), 2455–2458. https://doi.org/10.1039/C8GC00553B. (b) Allais, F.; Bonneau, G.; Peru, A.; Flourat, A. Method for Converting Levoglucosenone Intro 4-Hydroxymethyl Butyrolactone and 4-Hydroxymethyl Butenolide Without Using Any Organic Solvent and Catalyst. US201716315143, July 11, 2019.

64. (a) Mouterde, L. M. M.; Couvreur, J.; Langlait, M. M. J.; Brunois, F.; Allais, F. Identification and Expression of a CHMO from the *Pseudomonas aeruginosa* Strain Pa1242: Application to the Bioconversion of Cyrene™ into a Key Precursor (S)-γ-Hydroxymethyl-Butyrolactone. *Green Chem.* 2021, *23*(7), 2694–2702. https://doi.org/10.1039/D0GC04321D. (b) Allais, F.; Mouterde, L.; Stewart, J. D. Biocatalytic Method for Producing 2H-HBO and β-Susbtituted Analogues From LGO Using Cyclohexanone Monooxygenase. US2021388403, December 16, 2021.

65. (a) Peru, A. A. M.; Flourat, A. L.; Gunawan, C.; Raverty, W.; Jevric, M.; Greatrex, B. W.; Allais, F. Chemo-Enzymatic Synthesis of Chiral Epoxides Ethyl and Methyl (S)-3-(Oxiran-2-Yl)Propanoates from Renewable Levoglucosenone: An Access to Enantiopure (S)-Dairy Lactone. *Molecules* 2016, *21*(8), 988. https://doi.org/10.3390 /molecules21080988. (b) Allais, F.; Flourat, A.; Greatrex, B.; Raverty, W.; Duncan, A. Method for Synthesizing A Precursor of A Single Dairy-Lactone Isomer. p US10183923, January 22, 2019.

66. Moreaux, M.; Bonneau, G.; Peru, A.; Brunissen, F.; Janvier, M.; Haudrechy, A.; Allais, F. High-Yielding Diastereoselective Syn-Dihydroxylation of Protected HBO: An Access to D-(+)-Ribono-1,4-Lactone and 5-O-Protected Analogues. *Eur. J. Org. Chem.* 2019, *7*(7), 1600–1604. https://doi.org/10.1002/ejoc.201801780.

67. Diot-Néant, F.; Mouterde, L. M. M.; Couvreur, J.; Brunois, F.; Miller, S. A.; Allais, F. Green Synthesis of 2-Deoxy-D-Ribonolactone from Cellulose-Derived Levoglucosenone (LGO): A Promising Monomer for Novel Bio-based Polyesters. *Eur. Polym. J.* 2021, *159*, 110745. https://doi.org/10.1016/j.eurpolymj.2021.110745.

68. Shafizadeh, F.; Furneaux, R. H.; Pang, D.; Stevenson, T. T. Base-Catalyzed Oligomerization of Levoglucosenone. *Carbohydr. Res.* 1982, *100*(1), 303–313. https://doi.org/10.1016/S0008-6215(00)81044-7.

69. Corsaro, C.; Neri, G.; Santoro, A.; Fazio, E. Acrylate and Methacrylate Polymers' Applications: Second Life with Inexpensive and Sustainable Recycling Approaches. *Materials (Basel)* 2022, *15*(1), 282. https://doi.org/10.3390/ma15010282.

70. Mahboub, M. J. D.; Dubois, J.-L.; Cavani, F.; Rostamizadeh, M.; Patience, G. S. Catalysis for the Synthesis of Methacrylic Acid and Methyl Methacrylate. *Chem. Soc. Rev.* 2018, *47*(20), 7703–7738. https://doi.org/10.1039/C8CS00117K.

71. Wang, S.; Shuai, L.; Saha, B.; Vlachos, D. G.; Epps, T. H. From Tree to Tape: Direct Synthesis of Pressure Sensitive Adhesives from Depolymerized Raw Lignocellulosic Biomass. *ACS Cent. Sci.* 2018, *4*(6), 701–708. https://doi.org/10.1021/acscentsci .8b00140.

72. Bao, Y.; Ma, J.; Zhang, X.; Shi, C. Recent Advances in the Modification of Polyacrylate Latexes. *J. Mater. Sci.* 2015, *50*(21), 6839–6863. https://doi.org/10.1007 /s10853-015-9311-7.

73. Fouilloux, H.; Thomas, C. M. Production and Polymerization of Biobased Acrylates and Analogs. *Macromol. Rapid Commun.* 2021, *42*(3), 2000530. https://doi.org/10 .1002/marc.202000530.

74. Al-Naji, M.; Schlaad, H.; Antonietti, M. New (and Old) Monomers from Biorefineries to Make Polymer Chemistry More Sustainable. *Macromol. Rapid Commun.* 2021, *42*(3), 2000485. https://doi.org/10.1002/marc.202000485.

75. Mülhaupt, R. Green Polymer Chemistry and Bio-based Plastics: Dreams and Reality. *Macromol. Chem. Phys.* 2013, *214*(2), 159–174. https://doi.org/10.1002/ macp.201200439.

76. Ray, P.; Hughes, T.; Smith, C.; Simon, G. P.; Saito, K. Synthesis of Bioacrylic Polymers from Dihydro-5-Hydroxyl Furan-2-One (2H-HBO) by Free and Controlled Radical Polymerization. *ACS Omega* 2018, *3*(2), 2040–2048. https://doi.org/10.1021 /acsomega.7b01929.

77. Diot-Néant, F.; Rastoder, E.; Miller, S. A.; Allais, F. Chemo-Enzymatic Synthesis and Free Radical Polymerization of Renewable Acrylate Monomers from Cellulose-Based Lactones. *ACS Sustain. Chem. Eng.* 2018, *6*(12), 17284–17293. https://doi.org /10.1021/acssuschemeng.8b04707.

78. Kuo, S.-W.; Kao, H.-C.; Chang, F.-C. Thermal Behavior and Specific Interaction in High Glass Transition Temperature PMMA Copolymer. *Polymer* 2003, *44*(22), 6873–6882. https://doi.org/10.1016/j.polymer.2003.08.026.

79. Ray, P.; Hughes, T.; Smith, C.; Hibbert, M.; Saito, K.; Simon, G. P. Development of Bio-acrylic Polymers from Cyrene™: Transforming a Green Solvent to a Green Polymer. *Polym. Chem.* 2019, *10*(24), 3334–3341. https://doi.org/10.1039/C9PY00353C.

80. Hufendiek, A.; Lingier, S.; Prez, F. E. D. Thermoplastic Polyacetals: Chemistry from the Past for a Sustainable Future? *Polym. Chem.* 2018, *10*(1), 9–33. https://doi.org/10.1039/C8PY01219A.

81. Debsharma, T.; Behrendt, F. N.; Laschewsky, A.; Schlaad, H. Ring-Opening Metathesis Polymerization of Biomass-Derived Levoglucosenol. *Angew. Chem. Int. Ed. Engl.* 2019, *58*(20), 6718–6721. https://doi.org/10.1002/anie.201814501.

82. Debsharma, T.; Yagci, Y.; Schlaad, H. Cellulose-Derived Functional Polyacetal by Cationic Ring-Opening Polymerization of Levoglucosenyl Methyl Ether. *Angew. Chem. Int. Ed. Engl.* 2019, *58*(51), 18492–18495. https://doi.org/10.1002/anie.201908458.

83. Kaya, K.; Debsharma, T.; Schlaad, H.; Yagci, Y. Cellulose-Based Polyacetals by Direct and Sensitized Photocationic Ring-Opening Polymerization of Levoglucosenyl Methyl Ether. *Polym. Chem.* 2020, *11*(43), 6884–6889. https://doi.org/10.1039/D0PY01307B.

84. Debsharma, T.; Schmidt, B.; Laschewsky, A.; Schlaad, H. Ring-Opening Metathesis Polymerization of Unsaturated Carbohydrate Derivatives: Levoglucosenyl Alkyl Ethers. *Macromolecules* 2021, *54*(6), 2720–2728. https://doi.org/10.1021/acs.macromol.0c02821.

85. Hillmyer, M. A.; Laredo, W. R.; Grubbs, R. H. Ring-Opening Metathesis Polymerization of Functionalized Cyclooctenes by a Ruthenium-Based Metathesis Catalyst. *Macromolecules* 1995, *28*(18), 6311–6316. https://doi.org/10.1021/ma00122a043.

86. Isono, T.; Satoh, T. Poly(Cyclic Olefin)s. In: *Encyclopedia of Polymeric Nanomaterials*; Kobayashi, S., Müllen, K., Eds.; Springer: Berlin, Heidelberg, 2021, 1–8. https://doi.org/10.1007/978-3-642-36199-9_242-1.

87. Kohara, T. Development of New Cyclic Olefin Polymers for Optical Uses. *Macromol. Symp.* 1996, *101*(1), 571–579. https://doi.org/10.1002/masy.19961010163.

88. Nunes, P. S.; Ohlsson, P. D.; Ordeig, O.; Kutter, J. P. Cyclic Olefin Polymers: Emerging Materials for Lab-On-a-Chip Applications. *Microfluid. Nanofluid.* 2010, *9*(2), 145–161. https://doi.org/10.1007/s10404-010-0605-4.

89. Sabzekar, M.; Pourafshari Chenar, M.; Maghsoud, Z.; Mostaghisi, O.; García-Payo, M. C.; Khayet, M. Cyclic Olefin Polymer as a Novel Membrane Material for Membrane Distillation Applications. *J. Membr. Sci.* 2021, *621*, 118845. https://doi.org/10.1016/j.memsci.2020.118845.

90. Bruijns, B.; Veciana, A.; Tiggelaar, R.; Gardeniers, H. Cyclic Olefin Copolymer Microfluidic Devices for Forensic Applications. *Biosensors* 2019, *9*(3), 85. https://doi.org/10.3390/bios9030085.

91. Ward, D. D.; Shafizadeh, F. Cycloaddition "4+2" Reactions of Levoglucosenone. *Carbohydr. Res.* 1981, *95*(2), 155–176. https://doi.org/10.1016/S0008-6215(00)85573-1.

92. Bhaté, P.; Horton, D. Stereoselective Synthesis of Functionalized Carbocycles by Cycloaddition to Levoglucosenone. *Carbohydr. Res.* 1983, *122*(2), 189–199. https://doi.org/10.1016/0008-6215(83)88330-X.

93. Banwell, M. G.; Liu, X.; Connal, L. A.; Gardiner, M. G. Synthesis of Functionally and Stereochemically Diverse Polymers via Ring-Opening Metathesis Polymerization of Derivatives of the Biomass-Derived Platform Molecule Levoglucosenone Produced at Industrial Scale. *Macromolecules* 2020, *53*(13), 5308–5314. https://doi.org/10.1021/acs.macromol.0c01305.

94. Fadlallah, S.; Peru, A. A. M.; Flourat, A. L.; Allais, F. A Straightforward Access to Functionalizable Polymers through Ring-Opening Metathesis Polymerization of Levoglucosenone-Derived Monomers. *Eur. Polym. J.* 2020, *138*, 109980. https://doi.org/10.1016/j.eurpolymj.2020.109980.

95. Shafizadeh, F.; Essig, M. G.; Ward, D. D. Additional Reactions of Levoglucosenone. *Carbohydr. Res.* 1983, *114*(1), 71–82. https://doi.org/10.1016/0008-6215(83)88174-9.

96. Davydova, A. N.; Pershin, A. A.; Sharipov, B. T.; Valeev, F. A. Synthesis of Chiral 2,3-cis-Fused Butan-4-Olides from Levoglucosenone–1,3-Diene Diels–Alder Adducts. *Mendeleev Commun.* 2015, *25*(4), 271–272. https://doi.org/10.1016/j.mencom.2015.07.013.

97. Sebrão, D.; Sá, M. M.; Nascimento, M. da G. Regioselective Acylation of D-Ribono-1,4-Lactone Catalyzed by Lipases. *Process Biochem.* 2011, *46*(2), 551–556. https://doi.org/10.1016/j.procbio.2010.10.007.

98. Zhang, Q.; Song, M.; Xu, Y.; Wang, W.; Wang, Z.; Zhang, L. Bio-based Polyesters: Recent Progress and Future Prospects. *Prog. Polym. Sci.* 2021, *120*, 101430. https://doi.org/10.1016/j.progpolymsci.2021.101430.

99. *LIFE 3.0 - LIFE Project Public Page.* https://webgate.ec.europa.eu/life/publicWebsite/project/details/5731 (accessed 2022-07-08).

100. Diot-Néant, F.; Mouterde, L.; Fadlallah, S.; Miller, S. A.; Allais, F. Sustainable Synthesis and Polycondensation of Levoglucosenone-Cyrene-Based Bicyclic Diol Monomer: Access to Renewable Polyesters. *ChemSusChem* 2020, *13*(10), 2613–2620. https://doi.org/10.1002/cssc.202000680.

101. Kayishaer, A.; Fadlallah, S.; Mouterde, L. M. M.; Peru, A. A. M.; Werghi, Y.; Brunois, F.; Carboué, Q.; Lopez, M.; Allais, F. Unprecedented Biodegradable Cellulose-Derived Polyesters with Pendant Citronellol Moieties: From Monomer Synthesis to Enzymatic Degradation. *Molecules* 2021, *26*(24), 7672. https://doi.org/10.3390/molecules26247672.

102. Fadlallah, S.; Carboué, Q.; Mouterde, L. M. M.; Kayishaer, A.; Werghi, Y.; Peru, A. A. M.; Lopez, M.; Allais, F. Synthesis and Enzymatic Degradation of Sustainable Levoglucosenone-Derived Copolyesters with Renewable Citronellol Side Chains. *Polymers* 2022, *14*(10), 2082. https://doi.org/10.3390/polym14102082.

103. Shafizadeh, F.; Furneaux, R. H.; Stevenson, T. T. Some Reactions of Levoglucosenone. *Carbohydr. Res.* 1979, *71*(1), 169–191. https://doi.org/10.1016/S0008-6215(00)86069-3.

104. Matsumoto, K.; Ebata, T.; Koseki, K.; Okano, K.; Kawakami, H.; Matsushita, H. Short Synthesis of (3S,4R)- and (3R,4R)-3-Hydroxy-4-Hydroxymethyl-4-Butanolides, Two Lactones from Levoglucosenone. *Bull. Chem. Soc. Jpn.* 1995, *68*(2), 670–672. https://doi.org/10.1246/bcsj.68.670.

105. Kyriacos, D. Chapter 17 - Polycarbonates. In: *Brydson's Plastics Materials (Eighth Edition)*; Gilbert, M., Ed.; Butterworth-Heinemann, 2017, 457–485. https://doi.org/10.1016/B978-0-323-35824-8.00017-7.

106. Kumar, S.; Lively, B.; Sun, L. L.; Li, B.; Zhong, W. H. Highly Dispersed and Electrically Conductive Polycarbonate/Oxidized Carbon Nanofiber Composites for Electrostatic Dissipation Applications. *Carbon* 2010, *48*(13), 3846–3857. https://doi.org/10.1016/j.carbon.2010.06.050.

107. DeRudder, J. L. Commercial Applications of Polycarbonates. In: *Handbook of Polycarbonate Science and Technology*; chapter 4, 14 pages; Ed. John T. Bendler, CRC Press, Boca Raton, 2000.

108. Hellums, M. W.; Koros, W. J.; Husk, G. R.; Paul, D. R. Fluorinated Polycarbonates for Gas Separation Applications. *J. Membr. Sci.* 1989, *46*(1), 93–112. https://doi.org/10.1016/S0376-7388(00)81173-4.

109. Levchik, S. V.; Weil, E. D. Overview of Recent Developments in the Flame Retardancy of Polycarbonates. *Polym. Int.* 2005, *54*(7), 981–998. https://doi.org/10.1002/pi.1806.

110. Feng, J.; Zhuo, R.-X.; Zhang, X.-Z. Construction of Functional Aliphatic Polycarbonates for Biomedical Applications. *Prog. Polym. Sci.* 2012, *37*(2), 211–236. https://doi.org/10.1016/j.progpolymsci.2011.07.008.

111. *Polycarbonates Global and European Production 2016.* Statista. https://www.statista.com/statistics/650318/polycarbonates-production-worldwide-and-in-europe/ (accessed 2021-08-10).

112. Zhu, W.; Huang, X.; Li, C.; Xiao, Y.; Zhang, D.; Guan, G. High-Molecular-Weight Aliphatic Polycarbonates by Melt Polycondensation of Dimethyl Carbonate and Aliphatic Diols: Synthesis and Characterization. *Polym. Int.* 2011, *60*(7), 1060–1067. https://doi.org/10.1002/pi.3043.

113. Fadlallah, S.; Kayishaer, A.; Annatelli, M.; Mouterde, L. M. M.; Peru, A. A. M.; Aricò, F.; Allais, F. Fully Renewable Photocrosslinkable Polycarbonates from Cellulose-Derived Monomers. *Green Chem.* 2022, *24*(7), 2871–2881. https://doi.org/10.1039/D1GC04755H.

114. de Almeida, R. M.; Li, J.; Nederlof, C.; O'Connor, P.; Makkee, M.; Moulijn, J. A. Cellulose Conversion to Isosorbide in Molten Salt Hydrate Media. *ChemSusChem* 2010, *3*(3), 325–328. https://doi.org/10.1002/cssc.200900260.

115. Climent, M. J.; Corma, A.; Iborra, S. Converting Carbohydrates to Bulk Chemicals and Fine Chemicals over Heterogeneous Catalysts. *Green Chem.* 2011, *13*(3), 520–540. https://doi.org/10.1039/C0GC00639D.

116. Aricò, F.; Tundo, P. Isosorbide and Dimethyl Carbonate: A Green Match. *Beilstein J. Org. Chem.* 2016, *12*(1), 2256–2266. https://doi.org/10.3762/bjoc.12.218.

117. Aricò, F. Isosorbide as Biobased Platform Chemical: Recent Advances. *Curr. Opin. Green Sustain. Chem.* 2020, *21*, 82–88. https://doi.org/10.1016/j.cogsc.2020.02.002.

118. Zhang, M.; Lai, W.; Su, L.; Wu, G. Effect of Catalyst on the Molecular Structure and Thermal Properties of Isosorbide Polycarbonates. *Ind. Eng. Chem. Res.* 2018, *57*(14), 4824–4831. https://doi.org/10.1021/acs.iecr.8b00241.

119. Zhang, Z.; Xu, F.; He, H.; Ding, W.; Fang, W.; Sun, W.; Li, Z.; Zhang, S. Synthesis of High-Molecular Weight Isosorbide-Based Polycarbonates through Efficient Activation of Endo-Hydroxyl Groups by an Ionic Liquid. *Green Chem.* 2019, *21*(14), 3891–3901. https://doi.org/10.1039/C9GC01500K.

120. Vilela, C.; Sousa, A. F.; Fonseca, A. C.; Serra, A. C.; Coelho, J. F. J.; Freire, C. S. R.; Silvestre, A. J. D. The Quest for Sustainable Polyesters – Insights into the Future. *Polym. Chem.* 2014, *5*(9), 3119–3141. https://doi.org/10.1039/C3PY01213A.

121. Zenner, M. D.; Xia, Y.; Chen, J. S.; Kessler, M. R. Polyurethanes from Isosorbide-Based Diisocyanates. *ChemSusChem* 2013, *6*(7), 1182–1185. https://doi.org/10.1002/cssc.201300126.

122. Fadlallah, S.; Flourat, A. L.; Mouterde, L. M. M.; Annatelli, M.; Peru, A. A. M.; Gallos, A.; Aricò, F.; Allais, F. Sustainable Hyperbranched Functional Materials via Green Polymerization of Readily Accessible Levoglucosenone-Derived Monomers. *Macromol. Rapid Commun.* 2021, 2100284. https://doi.org/10.1002/marc.202100284.

8 Green Nanotechnology in Clean Water and Sanitation (SDG 6)

Mahak Kushwaha, Swati Agarwal, and Suphiya Khan

CONTENTS

8.1 INTRODUCTION

Humans have been battling and looking for nourishing food, pure drinking water, energy in a new form, and good health for more than a billion years. Over 1.7 billion people live in river basins throughout the world where water demand surpasses recharge at the moment. Likewise, more than 2.4 billion people lack access to even the most basic sanitation systems. It is difficult to significantly reduce the

DOI: 10.1201/9781003301769-8

amount of untreated wastewater while simultaneously boosting recycling and safe reuse on a worldwide scale (Aithal and Aithal, 2018). In order to manage water scarcity and thereby reduce the number of people impacted by it, it is challenging to increase the efficiency of water usage across all sectors while still maintaining a sustainable supply of drinkable water (Howard, 2021). Additionally, one-third of the global population lacks access to clean water and soap at home, making them particularly susceptible to the spread of infections like the COVID-19 virus. SDG 6 is focused on providing access to clean water and sanitation. It strives to achieve quality and sustainability of water resources globally as well as the availability and sustainable management of water and sanitation for everyone. It aims to assist billions of people throughout the world to live better lives by working in tandem with other sustainable development goals (SDGs).

While many economic sectors are using different nanoparticles (NPs) more often, there is rising interest in the biological and environmental safety associated with their production (Aithal and Aithal, 2021a). While avoiding any related toxicity, green nanotechnology offers techniques for converting biological systems to greenways for nanomaterial (NM) creation (Hashem, 2014). The use of noble metal nanoparticles in recent attempts to remove and identify such compounds at ultralow concentrations is outlined. An in-depth case study of the removal of pesticides using noble metal nanoparticles highlights significant difficulties encountered during the marketing of nano-based goods (Prabhakar et al., 2013). Due to the strong reactivity of nanoparticles, nanoremediation has become one of the most effective developing solutions for removing water pollutants (NMs).

8.2   CLEAN WATER AND SANITATION

For the protection of human well-being, particularly that of children, safe drinking water is crucial. Drinking water contains several pollutants, some of which are mentioned in Table 8.1. The majority of illnesses and fatalities among the underprivileged in developing nations are caused by water-related disorders (Hashem, 2014). According to the World Health Organization (WHO), infectious diarrhea accounts for the majority of water-related diseases (6% of all diseases globally; 70% or 1.7 million fatalities annually) (Kunduru et al., 2017). More than 4,500 kids under the age of five pass away every day from illnesses like diarrhea. Lack of access to clean water has a variety of negative repercussions in addition to sickness, issues with starvation, and problems with dehydration, which are as follows:

a. Having access to clean water close to home will free up time that can be used for work and education, which are the cornerstones of economic development.

b. Lack of access to water lowers the quantity and quality of food produced by both crops and cattle (Howard, 2021).

c. A lack of access to clean water worsens gender inequality and limits the opportunity for girls and women to receive an education, become literate, and engage in jobs that pay a living wage.

TABLE 8.1
Common Contaminants Classification in Waste and Drinking Water

Water Contaminants

Heavy Metals	Dyes	Microbes	Phenols	Antibiotics	Other
Cobalt (Co)	Azo	Viral	Chlorophenol	Macrolides	Oil agents
Copper (Cu)	Anthraquinone	Bacterial	Bromophenol	Tetracyclines	Fiber
Cadmium (Cd)	Sulfur-based	Protozoan	Nitrophenol	Fluoroquinolones	Plastics
Iron (Fe)	Phthalocyanine	Fungal	Alkylphenol	Sulfonamides	Organic matter
Lead (Pb)	Triarylmethane			Chloramphenicol	Inorganic matter
Arsenic (As)					Ions
Antimony (Sb)					Acid and alkali
Chromium (Cr)					
Selenium (Se)					
Mercury (Hg)					
Nickel (Ni)					
Zinc (Zn)					

  d. Due to habitat loss and landscape erosion, such as deserts, and dimin-
     ished, a lack of access to water can have an impact on the ecosystem and
     climate change.

One of the most important aspects of addressing issues of poverty, equity, and
related issues is sustainable water management (Khan, 2020). Rural regions are
home to 636 million people, or 83% of the world's population, who lack access
to better drinking water sources. In addition, concerns about the reliability and
security of several upgraded sources of drinking water continue. Therefore, the
actual number of individuals without access to clean drinking water may be two
to three times greater (Aithal and Aithal, 2018). A global problem, the lack of
clean drinking water, is expected to worsen as a result of population expansion
and environmental change. It is necessary to advance our economic and technical
current wastewater treatments and infrastructures to ensure a safe water supply
equally across all countries (Axon and James, 2018). At a consumption level, it is
needed to tackle the fact that global water use has increased at more than twice
the rate of population growth over the past century.

The world's most vulnerable nations to water scarcity are developing nations.
Freshwater availability and our capacity to get it are becoming less predictable
as climate change intensifies (Tortajada and Biswas, 2018). This is an even bigger problem in places where the public infrastructure is insufficient to meet the
growing demand for fresh water from the people (Chen et al., 2020). According to
the UN, there is already a problem with water shortage on every continent, both
because governments lack the necessary infrastructure and because there isn't
enough water.

Although still efficient, traditional water treatment techniques are no longer used. They are pricey and chemically intensive. As technology develops and we discover new methods to purify our water, those chemicals might not be required (Hambrey, 2017). Regardless of whether freshwater sources are used for human consumption in the future, new technology may be able to remove toxins from them or speed up the process of purifying drinking water. This might make it possible for developing nations to abandon outmoded technologies like fossil fuels in favor of more environmentally friendly ones (Howard, 2021). As a science-based application, technology offers tools to resolve issues and facilities to lead comfortable and enjoyable lives in society. One of the most fascinating of these is nanotechnology (Kruawal et al., 2005). As a result, new solutions for providing clean drinking water must be researched and explored. The use of nanotechnology is one strategy being researched in numerous nations to address the issue of increasing access to clean drinking water.

8.3   SDGS AND SDG 6

The United Nations' Sustainable development goals (SDGs) address the global challenges emerging in the 21st century, the 17 goals of UN sustainable development are interconnected (Jarvis, 2020). Especially, SDGs 3 and 6 on clean water and sanitation, Goal 3 on excellent health and well-being, Goals 12 and 13 on responsible consumption and production, as well as Goals 7 and 13 on cheap and clean energy call for new and creative industrial development techniques (Jarvis, 2020).

The study of green synthetic processes, green solvents, green extraction techniques, and a catalytic process invented by a green chemist is the very essence of the purpose to fulfil the sustainability goals for the world of the industrial economy (SDG, 2019). The SDGs are comprehensive goals set by the UN with the intent of eradicating poverty, preserving the environment, and ensuring the prosperity of all people. Clean water and sanitation, ethical production and consumption, and access to cheap, clean energy are just a few of the 17 sustainable development goals (SDGs) that will benefit greatly from environmental technology (Aithal and Aithal, 2018). The SDGs' goal 6 is related to access to clean water and sanitation. This objective seeks to guarantee that everyone has access to appropriate and equitable sanitation, among other significant and ambitious aims.

Many of the freshwater lakes and rivers of India, a nation with 16% of the world's population but just 4% of its water resources, are highly contaminated (Jarvis, 2020). The "bad" to the very poor water quality of the lake and the rivers that feed it is blamed for the extensive dumping of municipal trash, discharge of untreated residential wastewater, and agricultural run-off (Popa and Volf, 2006). We will need to create new materials that can effectively remove large amounts of contaminants from water and polluted soil while also being affordable, selectively recoverable, and reusable (Gurtu and Sharma, 2021). Environmental technology research has published a number of creative and extremely effective examples of such materials. This includes the creation of modified magnetic graphene oxide adsorbents to remove the herbicide glyphosate from water, adsorbents made from waste materials to remove the heavy metal Pb, and algae as a biosorbent to remove Zn from industrial effluent.

8.4   GREEN NANOTECHNOLOGY

New nanotechnology is predicted to meet both human beings' fundamental necessities and comfort demands. Humans have seven essential needs: food, water to drink, energy, clothing, housing, health, and the environment (Clark and Macquarrie, 2008). Green nanotechnology has the potential to offer practical, affordable, and ecologically responsible methods of delivering clean water for industrial and agricultural usage as well as drinkable water for human use (Guo et al., 2020). By supplying everyone with sustainable drinking water, nanotechnology breakthroughs in low-cost water filtration are anticipated to address the global drinking water crisis, making them a green technology.

One of the most valuable natural resources on Earth is water. It is mostly made of seawater. Only 3% of the world's freshwater supply is available, and two-thirds of it is frozen in glaciers, ice caps, and icebergs. The remaining 1% can be consumed by people (Aithal and Aithal, 2016). Currently, 1.1 billion people lack access to clean water, and 2.4 billion lack proper sanitation. Water-related illnesses are thought to be the cause of 3.4 million deaths annually, especially in children, in underdeveloped countries (Aithal and Aithal, 2021a). Freshwater is in greater demand. Currently, 70% of the water in the globe is used for agriculture. Demand for water will grow by 60% by 2030 to feed an additional 2 billion people. By 2050, over two-thirds of the world's population will be impacted by droughts due to current consumption, population, and development trends.

This problem will be resolved by green technology such as nanotechnology through costly decentralized water treatment, identification of impurities at the molecular level, and much-enhanced filtering systems (Aithal and Aithal, 2021b). This facilitates the large-scale, low-cost conversion of saltwater into potable water and the recycling of rainfall into clean drinking water. A large amount of drinking water may be produced using green nanotechnology-based water purification systems utilizing renewable solar or wind energy, making them sustainable and low maintenance.

8.5   NANOMATERIALS AS REMEDIATION AGENTS

8.5.1   NANOPARTICLES: COMMON CONTAMINANTS OF CONCERN IN SOIL AND GROUNDWATER

Depending on their form, size, and chemical characteristics, nanoparticles can be generally classified into a number of groups. The following list includes some of the well-known classes of NPs based on their physical and chemical features (Anjum et al., 2019).

8.5.1.1   Carbon Nanoparticles

Two important kinds of carbon-based NPs are fullerenes and carbon nanotubes (CNTs). Nanomaterial composed of globular hollow cages, such as allotropic forms of carbon, is found in fullerenes. For a variety of commercial uses, including fillers, effective gas adsorbents for environmental remediation, and as a support medium for various inorganic and organic catalysts, they have also generated notable economic interest in nanocomposites.

8.5.1.2   Metal Nanoparticles

Pure precursors of the metal are used to create metal NPs. These NPs are often made of organic materials, and the term polymer nanoparticle (PNP) collective is used specifically for them in the literature (Pareek et al., 2017).

8.5.1.3   Polymeric Nanoparticles

Polymeric nanoparticles (PNPs) are tiny particles with a size between 1 and 1,000 nm with the ability to contain or have active substances surface-adsorbed onto the polymeric core (Cano et al., 2020). Both nanospheres and nanocapsules are referred to as "nanoparticles", and they differ from one another in terms of morphological form (Hu et al., 2014). The two types of nanoparticles—nanospheres and nanocapsules—are included under the umbrella term "nanoparticle".

8.5.1.4   Magnetic Nanoparticle

The development of magnetic nanoparticle-based nanocontainers allows for the efficient removal of a variety of water contaminants (Hussain et al., 2022). The most effective adsorbents for water remediation applications are magnetite nanoparticles, which when further modified with carbon-based substances including carbon nanotubes, graphene, calixarenes, cyclodextrin, polymers, and bio-based substances show improved malleability (Kumari et al., 2020). It has been briefly explained and shown how recently developed magnetite nanoparticle-based nanocontainers may be used to treat water.

8.5.2   NANOCLAY

Clay and clay minerals are highly significant naturally occurring minerals that are essential for preserving the ecosystem. These clay minerals have been utilized for cleaning up contaminated water as well as for the storage and disposal of dangerous compounds (Awasthi et al., 2019). Due to its distinct qualities and features, nanoclay has attracted increased interest in recent years. Many adsorption techniques employ nanoclay as a nanoadsorbent. For the purification of water, nanoclays have been proven to be a very effective and efficient property enhancer.

8.5.3   NANOTUBES

Carbon nanotubes (CNTs) are cylindrical macromolecules with carbon atoms organized in a hexagonal lattice in the partitions of the tubes. The ends of the tubes are capped with the help of a structure resembling half of a fullerene (Horváth et al., 2022). Based on the degree to which the carbon atoms in the CNT layers have hybridized, they are categorized (Liu et al., 2013). As a result, CNTs include both multi-walled carbon nanotubes and single-walled carbon nanotubes (SWCNTs) (MWCNTs), and CNTs are used in various water purifying techniques (Arora and Attri, 2020).

Drinking water purification has been harder over the last few years for four main reasons. First, as a result of industrialization and urbanization, various harmful substances of human origin were released into natural surface water bodies. Toxic pollutants, including heavy metals (Kruawal et al., 2005), persistent organics (Liu et al., 2009), surface water released persistent organics, and endocrine disruptors ultimately end up in water treatment facilities. Second, pathogen removal from raw waters is significantly impacted by the dynamic character of the drinking water treatment process.

8.5.4 Nanofibers

Nanofibers range in size from 50 nm to 1,000 nm, and they have a lot of surface area, a lot of porosity, tiny pores, and low density. Molecular assembly, thermally induced phase separation, and electrospinning are several methods for creating nanofibers. Nanofibers are fiber-shaped, one-dimensional nanomaterials with diameters ranging from tens to hundreds of nanometers. In terms of their high surface area to volume ratio, linked nanoporosity, and superior mass transport capabilities, nanofibers are special materials. Nanofibers are fiber-shaped, one-dimensional materials having manometer-sized diameters. (10^{-9} m) (Tijing et al., 2019). These distinctive characteristics of these nanofibers include a high surface-to-volume ratio that provides a huge surface area, nanoporosity, and mass transport capabilities. Additionally, a vast array of organic, synthetic, and hybrid polymers with different physical, chemical, and mechanical characteristics may be used to create nanofiber (Correa et al., 2019). These special attributes provide them exceptional water permeability, the superior mechanical strength needed for filtering, selective adsorption to substances such as hydrophobic phenols and CO_2, low polarization resistance even at high pressure, and other special qualities.

Application of nanofibers in desalination, water, and wastewater treatment.

a. Nanofibers for water treatment and desalination.
b. Membrane support layer for forward osmosis using nanofibers (FO).
c. Nanofibers as capacitive deionization electrodes (CDI).
d. Reverse osmosis using nanofibers as a barrier/mid-layer (RO).
e. Solar steam generation using a porous floating membrane made of nanofibers.
f. Using nanofibers as filters or adsorbents to remove contaminants from water.

Pump and treat (P&T) and in-situ remediation methods are the two categories under which groundwater remediation technologies are categorized. Pumped out of the aquifer and treated externally in a treatment facility, polluted groundwater is used for

P&T. Therefore, groundwater remediation may make use of all surface water treatments, including adsorption, filtration, and advanced oxidation processes (AOPs). The use of nanofibers as the primary or substrate material for the aforementioned treatment methods that may be used for P&T groundwater remediation is now also a possibility as a result of this. Based on its major function in the treatment process, nanofiber can be used in one of two ways for groundwater remediation. One method is to purge groundwater of pollutants by using nanofiber itself. The other is to use the nanofiber as a substrate or carrier to help groundwater remediation methods already in use become immobilized. Because of their large specific surface area and simple surface functionalization, nanofibers, for instance, have been employed in adsorption and filtering (Haider et al., 2015). The primary contaminants in groundwater that are often remedied by the adsorption method are heavy metals (such as mercury, copper, cadmium, lead, arsenic, arsenic, and chromium) and organic pollutants (such as chloride organics, dyes). Adsorption primarily functions through affinities between pollutants and functional groups on adsorbents (such as physical affinity, electrostatic contact, chemical chelation, and complexation).

8.5.5 NANOMEMBRANES

Modern research is focused on developing sustainable and affordable membrane technology (MT) for the purification of water. Due to their exceptional properties, two-dimensional (2D), material-based membranes made of graphene, metal-organic frameworks (MOFs), zeolite, transition metal dichalcogenides (TMDCs), MXenes, graphitic carbon nitride (g-C3N4), and hexagonal boron nitride (hBN) have shown tremendous potential for use in a variety of separation processes. Nanotechnology has the potential to enhance the quality, accessibility, and long-term viability of water resources. Numerous membrane types, such as nanofiltration (NF), reverse osmosis (RO), microfiltration (MF), and ultrafiltration, have been used for water purification (UF). Although RO has the capacity to produce water with the highest purity, nanofiltration membrane has emerged as a modern method for water purification. If water is allowed to pass through ultrafiltration (UF) membranes, the macro-size molecules and colloids are removed. Ultrafiltration membranes have pores that range in size from 2 to 100 nm (Tlili and Alkanhal, 2019). Membranes used in nanofiltration are incredibly economical when compared to alternative filtering techniques. Additionally, in addition to total dissolved solids, NF membranes can efficiently remove a wide range of salts, minerals, pathogens (fungi, molds, viruses, and bacteria), monovalent and multivalent cations, anions, and other suspended nanoparticles from surface and groundwater (TDS). The NF membrane has several engineering and industrial uses, including those in oil, textiles, drinks, food, chemicals, and a wide range of other industries. It is generally known that the pore size of nanofiltration membranes is typically quite small—around 1 nm—helping to separate bigger molecules from smaller ones as well as assisting in the elimination of microorganisms. Utilizing the contact angle, the modified cellulose and pH wettability of the membrane through mineral oil and water are assessed.

Numerous researchers have looked into the usage of nonreactive membranes composed of metal nanoparticles and nanostructured membranes made of nano-materials such nanoparticles, dendrimers, and carbon nanotubes. Absorption is a popular approach for water filtration because of its efficiency, adaptability, and reasonably low process costs (Feng et al., 2013). Operative absorbents that remove various pollutants from contaminated water include activated carbon, modified clays, zeolites, silica, and layered double hydroxides. Nanotechnology offers a revolutionary method for safe water supply, filtration, and distribution.

8.6 NANOREMEDIATION PROCESS

The main approaches to remediation are known as ex situ and in situ.

8.6.1 *Ex situ*

Ex situ refers to physically removing the contaminated matrix, such as by digging up soil or pumping groundwater and then having it treated elsewhere. Ex-situ therapies, on the other hand, are slower and more expensive but are easier to monitor (Hussain et al., 2022). Ex-situ cleanup methods are anticipated to gradu-ally disappear in favor of in-situ remediation methods in the future.

8.6.2 *In situ*

Permeable reactive barriers (PRBs) are a typical in-situ, or below-ground, reme-diation technique now utilized to clean up contaminated groundwater. PRBs are treatment areas made of substances that contaminate degradation or immobiliza-tion when groundwater travels over the barrier (Marcon et al., 2021). The limita-tion of PRBs' remediation capabilities to pollutant plumes that flow through them prevents them from addressing densely contaminated groundwater that is located outside of the barrier and involves treating pollutants on-site. The fundamental advantage of in-situ procedures is that contaminated material does not need to be removed or transported, thus reducing the related expenses. Physical, chemical, or biological techniques can all be used to achieve in-situ environmental rehabili-tation. Thermotherapy (steam-enhanced extraction, electrical resistive heating, or thermal conductive heating), chemical oxidation, surfactant cosolvent flushing, and bioremediation are examples of in-situ treatment methods.

8.7 APPLICATION OF GREEN NANOTECHNOLOGY IN CLEAN WATER

Following are a few uses for nanotechnology that were once considered environ-mentally friendly:

a. Clean, safe, cost-effective, renewable energy.
b. Recyclable materials that are stronger, lighter, and more robust.

c. Inexpensive filters to purify saltwater for use as drinking water.

d. Medical tools and medications that can diagnose and cure illness more quickly and with minimal or no negative effects (Dhage and Shisodiya, 2013).

e. Lighting that consumes a small portion of the energy required by traditional systems.

f. Environmental cleanup tools and sensors that can discover and recognize harmful chemical and biological pollutants.

g. Sustainable infrastructure and green construction (Dhage and Shisodiya, 2013).

h. Production methods that have been altered to reduce greenhouse gas emissions.

8.8 LIMITATIONS AND FUTURE ASPECTS OF GREEN NANOTECHNOLOGY

Although there are several advantages to nanotechnology, such as enhanced industrial processes, better environmental conditions, and water purification systems are effective renewable energy systems (Gurtu and Sharma, 2021). Improvements in the properties and functionality of physical systems, with optimization of health issues through nanomedicine, improved food production techniques and improved nutrition in food (Aithal and Aithal, 2016). If large-scale infrastructure autofabrication using self-replicating machines, etc., is not managed correctly with the creation of suitable awareness and safeguards, it might have negative effects on the health of living things, environment, social life, and economics of the country. Several of the problems are connected to health, the environment, society, and the economy.

Due to its applicability in every aspect of society, green nanotechnology will develop into a multipurpose technology (Dhage and Shisodiya, 2013). It will have a significant impact on virtually all industries and aspects of society since it will offer cleaner, safer, and smarter products for the home, communications, health, transportation, agriculture, and industry in general in its evolved form. So, by properly applying nanotechnology for environmental sustainability, it may be developed as a green technology for a sustainable civilization.

REFERENCES

Aithal, S., & Aithal, P. S. (2016). Opportunities & challenges for green technology in the 21st century. *International Journal of Current Research and Modern Education (IJCRME), 1*(1), 818–828.

Aithal, S., & Aithal, P. S. (2018). Concept of ideal water purifier system to produce potable water and its realization opportunities using nanotechnology. *International Journal of Applied Engineering and Management Letters (IJAEML), 2*(2), 8–26.

Aithal, S., & Aithal, P. S. (2021a). Green and eco-friendly nanotechnology–concepts and industrial prospects. *International Journal of Management, Technology, and Social Sciences (IJMTS), 6*(1), 1–31.

Aithal, S., & Aithal, P. S. (2021b). Green nanotechnology innovations to realize UN sustainable development goals 2030. *International Journal of Applied Engineering and Management Letters (IJAEML)*, *5*(2), 96–105.

Anjum, M., Miandad, R., Waqas, M., Gehany, F., & Barakat, M. A. (2019). Remediation of wastewater using various nano-materials. *Arabian Journal of Chemistry*, *12*(8), 4897–4919.

Arora, B., & Attri, P. (2020). Carbon nanotubes (CNTs): A potential nanomaterial for water purification. *Journal of Composites Science*, *4*(3), 135.

Awasthi, A., Jadhao, P., & Kumari, K. (2019). Clay nano-adsorbent: Structures, applications, and mechanism for water treatment. *SN Applied Sciences*, *1*(9), 1–21.

Axon, S., & James, D. (2018). The UN Sustainable Development Goals: How can sustainable chemistry contribute? A view from the chemical industry. *Current Opinion in Green and Sustainable Chemistry*, *13*, 140–145.

Cano, A., Sánchez-López, E., Ettcheto, M., López-Machado, A., Espina, M., Souto, E. B., … Turowski, P. (2020). Current advances in the development of novel polymeric nanoparticles for the treatment of neurodegenerative diseases. *Nanomedicine*, *15*(12), 1239–1261.

Chen, T. L., Kim, H., Pan, S. Y., Tseng, P. C., Lin, Y. P., & Chiang, P. C. (2020). Implementation of green chemistry principles in the circular economy system towards sustainable development goals: Challenges and perspectives. *Science of the Total Environment*, *716*, 136998.

Clark, J. H., & Macquarrie, D. J. (Eds.). (2008). *Handbook of green chemistry and technology*. John Wiley & Sons.

Correa, D. S., Mercante, L. A., Schneider, R., Facure, M. H., & Locilento, D. A. (2019). Composite nanofibers for removing water pollutants: Fabrication techniques. In *Handbook of eco-materials* (pp. 441–468). Springer-Nature.

Dhage, S. D., & Shisodiya, K. K. (2013). Applications of green chemistry in sustainable development. *International Research Journal of Pharmacy*, *4*(7), 1–4.

Feng, C., Khulbe, K. C., Matsuura, T., Tabe, S., & Ismail, A. F. (2013). Preparation and characterization of electrospun nanofiber membranes and their possible applications in water treatment. *Separation and Purification Technology*, *102*, 118–135.

Guo, M., Nowakowska-Grunt, J., Gorbanyov, V., & Egorova, M. (2020). Green technology and sustainable development: Assessment and green growth frameworks. *Sustainability*, *12*(16), 6571.

Gurtu, M. G., & Sharma, M. (2021). Major applications of green nanotechnology in Sustainable development and its impact on the global environment: A review. *Nveo-Natural Volatiles & Essential Oils Journal NVEO*, *8*(4), 12823–12851.

Haider, S., Haider, A., Ahmad, A., Khan, S. U. D., Almasry, W. A., & Sarfarz, M. (2015). Electrospun nanofibers affinity membranes for water hazards remediation. *Nanotechnology Research Journal*, *8*(4), 511.

Hambrey, J. (2017). The 2030 agenda and the sustainable development goals: The challenge for aquaculture development and management. *FAO Fisheries and Aquaculture Circular*, *C1141*.

Hashem, E. A. (2014). Nanotechnology in water treatment, case study: Egypt. *Journal of Economics and Development Studies*, *2*(3), 243–259.

Horváth, E., Gabathuler, J., Bourdiec, G., Vidal-Revel, E., Benthem Muñiz, M., Gaal, M., … Forró, L. (2022). Solar water purification with photocatalytic nanocomposite filter based on TiO_2 nanowires and carbon nanotubes. *npj Clean Water*, *5*(1), 1–11.

Howard, G. (2021). The future of water and sanitation: Global challenges and the need for greater ambition. *Journal of Water Supply: Research and Technology-Aqua*, *70*(4), 438–448.

Hu, X., Liu, S., Zhou, G., Huang, Y., Xie, Z., & Jing, X. (2014). Electrospinning of polymeric nanofibers for drug delivery applications. *Journal of Controlled Release*, *185*, 12–21.

Hussain, A., Rehman, F., Rafeeq, H., Waqas, M., Asghar, A., Afsheen, N., … Iqbal, H. M. (2022). In-situ, Ex-situ, and nano-remediation strategies to treat polluted soil, water, and air–A review. *Chemosphere*, *289*, 133252.

Jarvis, P. (2020). Environmental technology for the sustainable development goals (SDGs). *Environmental Technology*, *41*(17), 2155–2156.

Khan, S. H. (2020). Green nanotechnology for the environment and sustainable development. In Mu. Naushad, & Eric Lichtfouse (Eds.), *Green materials for wastewater treatment* (pp. 13–46). Cham: Springer.

Kruawal, K., Sacher, F., Werner, A., Müller, J., & Knepper, T. P. (2005). Chemical water quality in Thailand and its impacts on drinking water production in Thailand. *Science of the Total Environment*, *340*(1–3), 57–70.

Kumari, P., Kumar, S., & Singhal, A. (2020). Magnetic nanoparticle-based nanocontainers for water treatment. In Phuong Nguyen-Tri, Trong-On Do, & Tuan Anh Nguyen (Eds.), *Smart nanocontainers* (pp. 487–498). Amsterdam: Elsevier.

Kunduru, K. R., Nazarkovsky, M., Farah, S., Pawar, R. P., Basu, A., & Domb, A. J. (2017). Nanotechnology for water purification: Applications of nanotechnology methods in wastewater treatment. *Water Purification*, 33–74.

Liu, M., Sun, J., Sun, Y., Bock, C., & Chen, Q. (2009). Thickness-dependent mechanical properties of polydimethylsiloxane membranes. *Journal of Micromechanics and Microengineering*, *19*(3), 035028.

Liu, X., Wang, M., Zhang, S., & Pan, B. (2013). Application potential of carbon nanotubes in water treatment: A review. *Journal of Environmental Sciences*, *25*(7), 1263–1280.

Marcon, L., Oliveras, J., & Puntes, V. F. (2021). In situ nanoremediation of soils and groundwaters from the nanoparticle's standpoint: A review. *Science of the Total Environment*, *791*, 148324.

Pareek, V., Bhargava, A., Gupta, R., Jain, N., & Panwar, J. (2017). Synthesis and applications of noble metal nanoparticles: A review. *Advanced Science, Engineering and Medicine*, *9*(7), 527–544.

Popa, V., & Volf, I. (2006). Green chemistry and sustainable development. *Environmental Engineering and Management Journal (EEMJ)*, *5*(4), 545–558.

Prabhakar, V., Bibi, T., Vishnu, P., & Bibi, T. (2013). Nanotechnology, future tools for water remediation. *International Journal of Emerging Technology and Advanced Engineering*, *3*(7), 54–59.

SDG, U. (2019). Sustainable development goals. *The energy progress report. Tracking SDG, 7*.

Tijing, L. D., Yao, M., Ren, J., Park, C. H., Kim, C. S., & Shon, H. K. (2019). Nanofibers for water and wastewater treatment: Recent advances and developments. In Xuan-Thanh Bui, Chart Chiemchaisri, Takahiro Fujioka, & Sunita Varjani (Eds.), *Water and wastewater treatment technologies* (pp. 431–468). Springer-Singapore.

Tlili, I., & Alkanhal, T. A. (2019). Nanotechnology for water purification: Electrospun nanofibrous membrane in water and wastewater treatment. *Journal of Water Reuse and Desalination*, *9*(3), 232–248.

Tortajada, C., & Biswas, A. K. (2018). Achieving universal access to clean water and sanitation in an era of water scarcity: Strengthening contributions from academia. *Current Opinion in Environmental Sustainability*, *34*, 21–25.

9 Indian Knowledge System

The Roots of Environmental Sustainability, Civilization, and Green Chemistry

Sanjay K. Sharma and Komal Makhijani

CONTENTS

9.1 INTRODUCTION

Throughout the world, industrialization started in the 18th century owing mainly to increased population, exploration of resources, and demand for quality of life. In fact, industrial development is important for socioeconomic reasons as it not only provides the productive foundation for economic growth and prosperity of nations but also has a positive impact on multiple aspects of the social fabric such as poverty eradication, income inequality reduction, and participation in formal

DOI: 10.1201/9781003301769-9

219

and non-formal education. However, haphazard industrialization disrupts the ecosystem and poses serious consequences to the environment.

To maintain a balance between rapid industrialization and environment conservation, industries should incorporate new technologies and processes that make production more resource-efficient, decoupling growth from environmental damage. In this endeavor, the concept of "green chemistry" was introduced in the late 1990s which is 'the design, development, and implementation of chemical products and processes to reduce or eliminate the use of substances hazardous to human health and the environment' [1]. In the 1990s, Paul Anastas and John Warner postulated the 12 principles of green chemistry which propose environmentally favorable actions from the planning of a product to its synthesis, processing, analysis, and destination after use, thus minimizing environmental and occupational hazards [2].

Green chemistry principles are a set of guidelines and concepts aimed at reducing the negative impact of chemical processes and products on the environment and human health. These principles are in line with the traditional Indian knowledge system, which has long emphasized the importance of living in harmony with nature and using natural resources in a sustainable way.

In particular, the principles of green chemistry align with several key concepts in Indian knowledge systems, such as Ayurveda, which emphasizes the importance of using natural materials in medicine and wellness practices. Ayurveda also stresses the importance of understanding the interdependence of all living beings and the need to minimize harm to the environment in pursuit of human goals.

Other Indian knowledge systems, such as Yoga and Vastu Shastra, also promote holistic approaches to wellness and sustainable living that are in line with the principles of green chemistry. These systems encourage the use of natural materials, the reduction of waste, and the promotion of health and well-being through mindful and sustainable practices.

Overall, the principles of green chemistry and traditional Indian knowledge systems share a common goal of promoting sustainable living and reducing the negative impact of human activity on the environment. By applying the principles of green chemistry, we can draw on the wisdom of traditional Indian knowledge systems to create a more sustainable and harmonious world for all.

In India, industries which have may harm the environment such as chemical, iron and steel, sugar, cement, glass, and other consumer goods, began only after the middle of the 19th century. Although several environmental protection laws existed even before independence, the Ministry of Environment and Forest (MoEF) was set up in 1985 in collaboration with the Pollution Control Board's Environment, Air, Water, Solid, and Hazardous Waste Protection Acts.

Even before the formation and implementation of the regulatory framework in the late 1990s, literature shows that alchemists have always striven to protect the environment. The principles of sustainability were deeply rooted and practiced in the Vedic, Jain, Buddhist, and Kautilya's Arthshashtra eras. The ancient Indian literature clearly suggests the existence of ecological sustainability which is replicated in many modern holistic methods. Thus, the concept of green chemistry is not new in the Indian context. In this chapter, emphasis is given on (1) history, (2) significant presence of green chemistry principles in ancient India, (3) impact on the environment, and (4) challenges and possible solutions.

9.2 HISTORY

Indian literature from the 2nd century BC contains a growing amount of knowledge on chemistry that initially seemed unrelated. However, over time, these various pieces of knowledge became more interconnected and developed stronger technological connections with each other. This evolution is contributed to the human pursuit of longevity, wealth, and happiness. The eminent Indian chemist of the last century and a historian of chemistry, Acharya Prafulla Chandra Ray stated that the five stages in the development of chemistry are (1) the pre-Vedic stage upto 1500 BC, including the Harappan period, (2) the Vedic and the Ayurvedic period upto 700 CE, (3) the transitional period from 700 CE to 1100 CE, (4) the Tantric period from 700 CE to 1300 CE, and (5) the "Iatro-Chemical period" from 1300 CE to 1600 CE [3].

In ancient India, amazingly detailed information is available on the existence, exploration, and innovation of chemicals, metals, glass, dyes, cosmetics, perfumery, and herbal medicines through various methods and processes which were collectively called Rasayan Shastra, Rastantra, Raskriya, or Rasvidya [4]. The following section discusses the evolution of chemistry through the ages.

9.2.1 INDUS VALLEY CIVILIZATION

Archaeological findings at the Harrapan and Mohenjodaro sites prove the existence of a strategically developed society depicted in terms of streets, public baths, temples, and granaries. They also had the means of mass production of pottery, houses made of baked bricks, and a script of their own. Systematic excavations at Mohenjodaro and Harrapa substantiate remains of glazed pottery, use of gypsum in construction work, presence of ornaments made from glass, silver, and gold, and artifacts made from copper which were hardened using tin and arsenic. The civilization also supports evidence of technologies used for the extraction of copper and iron [5].

9.2.2 VEDIC PERIOD

Vedic chemistry mentions metals like gold, copper, silver, and bronze. Chemists during the Vedic period opined that these metals could only be obtained through complex chemical processes. Atharvaveda compares the color of the universal power with that of metals. The technique for metallurgical alloying is also presented in Chandogya Upanishad. Copper utensils; iron, gold, and silver ornaments; terracotta discs; and painted grey pottery have been found in many archaeological sites in north India. The golden gloss of the Northern Black Polished Ware made during 600–200 BC is still a chemical mystery that cannot be replicated. According to Rigveda, tanning of leather and dyeing of cotton was practiced during this period [6–9].

9.2.3 POST-VEDIC PERIOD

During this period, Indian chemistry flourished. Various schools of philosophy, especially Samkhya, Nyaya and Vaiseshika, Arthasastra of Kautilya, Brihat Samhita of Varahamihira, and Ayurvedic texts like Charaka Samhita and Susruta Samhita provided a lot of evidence regarding the advanced practices of Indian chemistry during this period [10].

Classical texts such as Brahmanas, Upanishads, and Puranas emphasize valuable information about the chemical activities during the Vedic era. Kautilya's Arthasastra (KA) which described the production of salt from the sea and the collection of shells, diamonds, pearls, and corals was a scientific landmark of this era. KA Charaka Samhita and Susruta Samhita are two celebrated Ayurvedic treatises on medicine and surgery; the chemical knowledge of these times especially that related to medicine was compiled in them.

Sushruta Samhita explains the importance of Alkalies whereas the Charaka Samhita mentions the preparation of sulfuric acid, nitric acid, and oxides of copper, tin, and zinc; the sulfates of copper, zinc, and iron; and the carbonates of lead and iron. Rasopanishada describes the preparation of gunpowder mixture. Chakrapani discovered mercury sulfide and also soap. Varähmihir's Brihat Samhita informs about the invention of glutinous material extracted from various plants, fruits, seeds, and barks which can be applied on walls and roofs of houses and temples. Apart from this, Brihat Samhita also contains references to perfumes and cosmetics. Even the process of fermentation is mentioned in most of these texts [11].

9.2.4 MEDIEVAL ALCHEMY

Indian alchemy mainly concentrated on gold making and elixir synthesis. Here, both the metallurgical and the physico-religious aspects were superimposed to get a divine material. The process involved the use of mercury and the elixirs made from it were used in the transmutation of the base metals into noble ones. When administered internally it claimed to purify and rejuvenate the body ultimately leading to an imperishable and immortal state. In the early 8th century, the rudiments of a separate discipline of Rasayana appeared when the great Buddhist exponent Nagarjuna rendered the knowledge into a sastra or theoretical form. The most scientific aspect of his work is contained in its observation of materials: dravya and dravya rasa.

There were also texts like Patanjali's Yogasutra with Vyasa Bhasya, which introduced an enlightened view of matter or "Bhuta" in the Sankhya tradition. The compilation of Charaka Samhita demonstrated extensive knowledge of "rasa" or the essence of botanical and zoological ingredients, as well as "rasayana" or the process of preparing drugs from these essences. Finally, the Kautilya Arthasastra discussed complex topics related to mining, identifying minerals, and identifying mineral deposits, which was the most significant among them.

So, the birth of "Rasayana" was the result of contributions from several working disciplines through the centuries. Nagarjuna rendered the format of a rational, theoretical, and analytical framework combining concepts of matter using a huge inventory of knowledge and information from practicing metallurgists, potters, physicians, cosmetologists, masons, and horticulturists.

In the early Indian Ayurveda parlance, Rasayana was related to the special treatment for longevity. In fact, the Rasayana section in the Ayurvedic treatises represents the parallel repertoire of the medical practitioners meant for the healthy contingent of patients who aspired to a long life [12, 13].

9.2.5 MODERN CHEMISTRY

A sharp decline in alchemical ideologies was observed from the beginning of the Tantric period with the realization that alchemy could not deliver what it promised. This led to the ascendance of iatrochemistry which probably reached a steady state over the next 150–200 years. However, it also started fading due to various reasons, including the introduction and practice of western medicine, the gap between practicing old Ayurvedic methods and reluctance to adopt new foreign ones, and the adverse policies of the rulers. Modern chemistry appeared on the Indian scene in the late 19th century when several young scientists, such as P.C. Ray, Chuni Lal Bose, Chandra Bhusan Bhaduri and Jyoti Bhusan Bhaduri, and R.D. Phookan, began to take a keen interest in modern scientific research activities and laid the foundation for inventions [5–9].

9.3 SIGNIFICANT PRESENCE OF "GREEN CHEMISTRY" IN ANCIENT INDIA

As people around the world strive for economic growth and a better quality of life, they are causing harm to the environment by overusing natural resources. This environmental degradation is a major issue, and it is important that we adopt the concept of "sustainability." Sustainability means taking responsibility for the long-term well-being of the environment, the economy, and society as a whole.

On the international level, the first concrete step was taken in 1992 at the United Nations Conference, Stockholm, where the conceptualization of environmental protection was elucidated in terms of 12 principles as depicted below-

1. Prevention of waste generation: It is better to avoid generating waste than to treat it after its generation.
2. Atomic economy: Synthetic methods should be planned so that the final product incorporates as much of the reagents used during the process as possible. Thus, waste generation will be minimized.

3. Safer chemical synthesis: Synthetic methods should be designed to use and generate substances with low or no occupational and environmental toxicity. Thus, the replacement of toxic solvents with low-toxicity or nontoxic solvents is highly recommended.

4. Safer chemicals design: Great importance should be given to the toxicity of the designed chemicals. They should obviously fulfill their functions, but should also present the lowest possible toxicity.

5. Use of safer solvents and auxiliaries: The use of solvents and other reagents should be avoided where possible. When it is not possible, these substances should be innocuous.

6. Design for energy efficiency: Choose the least energy-intensive route. Avoid heating and cooling or pressurized vacuum conditions unnecessarily.

7. Use of renewable raw materials: Whenever it is economically and technically feasible, renewable raw materials should be used instead of nonrenewable.

8. Reduction of derivatives: Unnecessary derivatization processes should be avoided or minimized, as they require the additional use of reagents and, therefore, generate waste.

9. Catalysis: The use of catalytic reagents (as selective as possible) is better than the use of stoichiometric reagents.

10. Degradation products design: Chemicals should be designed so that at the end of their function they decompose into harmless degradation products and do not persist in the environment.

11. Real-time analysis for pollution prevention: Analytical methods should be monitored in real time to avoid the formation of hazardous substances.

12. Accidents prevention: Both the substances and the way they are used in a chemical process should be chosen considering minimization of potential accidents, such as leaks, explosions, and fires, aiming at greater occupational and environmental safety [14].

The crux of the 12 principles is to minimize waste generation; maximum conversion of raw materials into products; design safer reactants, solvents, reagents, products, and processes; use renewable resources instead of nonrenewable; create energy efficient processes; and implement real-time monitoring of methods to avoid formation of hazardous materials, prevention of gas or chemical leaks, fires at sites leading to harmful impact on the environment. Thus environmental sustainability and ecology can be attained through the preservation, protection, and management of all the resources. The efforts are to allow economic progress and environmental growth to proceed in harmony [14].

Although ancient India did not face ecological problems, the awareness for preserving the geographical, climatic, and environmental conditions preventing pollution can surely be traced in our ancient literature. The Vedic, Jain, Buddhist, and Kautilya's Arthshashtra established the principles of sustainability centuries ago. It is true that no separate literary treatises that deal with the environment were in existence back then but indirect mention of these principles is found in literary records such as the Vedas, Smritis, Samhitas, Puranas, Upanishads, and Niti Shastras [15, 16].

These sacred books state the importance of safeguarding the environment and preventing pollution along with the liability of mankind towards nature through various verses, folklore, art culture, and religious practices that treasured solar energy, trees, and wildlife [17]. The Vedas, such as Rigveda, Yajurveda, Samaveda, and Atharvaveda, clearly recognize the importance of maintaining the seasonal cycles that lead to climatic changes largely due to inappropriate human behavior and actions. The ancients treated nature holistically giving utmost reverence to preserving its various entities and elements [15–17].

Hindu theology emphasizes that the elements of nature should be considered holy, which is practiced even in present times. The five cardinal elements— "earth," "water," "light," "cosmos," and "air"—were worshipped, safeguarded, and considered symbolic during ancient India. The Rigvedic hymns refer to many gods and goddesses that were the personification of natural entities like gods of the sun, moon, thunder, water, rivers, rain, lightning, trees, etc. They have been glorified and worshipped as givers of health, wealth, and prosperity. The idea of worshipping nature does not limit us to revering the deities, but also gives due dominance to nature over mankind [18].

The importance of the Sun and the related solar energy that regulates the water chain and food chain and drives various other nutrient cycles thereby establishing total control over the Earth's ecosystem was well understood and realized by the ancient Indians.

The Vedic Hymn to the Earth celebrated it as a "Mother" for all her natural bounties and particularly for her flora and fauna. Her blessings are sought for prosperity in all endeavors and fulfillment of all righteous aspirations [17].

9.3.1 FLORA AND FAUNA

The Manusmriti also shows an awareness of the importance of ecology by urging people to protect plants and animals. In ancient India, the forests around villages were classified into different types, including Sreevanam, Upavanam, and Tapovanam. Sreevanam, which means "forests of prosperity," were dense forests and groves that were located very close to the village. The law stated that people could use only a limited amount of forest products (e.g., dead wood, fruits, flowers, leaves, forest produce, etc.) required for their livelihood. To enrich the existing ecosystem people carried their duty of planting and nurturing trees in these groves. After the Sreevanam the next peripheral layer of forest is Mahavanam, a type of reserved forest with no human intervention. Tapovanam were the densest forest, where only the great saints could go for meditation. The saints associated trees with gods and goddesses thus making them sacred. In the great epic Ramayana, Rama's entire journey from Ayodhya to Lanka was through forests and sea. Similarly in the Mahabharata, the Pandavas spent their years of exile in the forest and made marriage alliances with forest tribes. The divinity was purposely attributed to forests for the maintenance of ecological balance. The faith in religion and fear of wrath for disrespecting gods restricted people from cutting trees unnecessarily. Thus, there was no problem of imbalance due to deforestation [19].

Even wild and domesticated animals were given due respect in the ancient Indian tradition. Many Hindu gods and goddesses have some particular animal or bird as their vehicles (vahana). The association of these animals with peoples' religious beliefs played a prominent role in their protection and conservation for so long in India.

Buddhism and Jainism, the two most important religions of India, also propagated the belief in complete non-violence (Ahimsa). They criticized the killing of animals in the name of sacrifice or rituals for balancing the life cycle and advocated a gentle and non-aggressive attitude towards nature and prudent use of resources. All the 24 Tirthankaras of Jainism were closely associated with the environment in one way or another [20].

According to Manusmriti, the highest level of environmental ethics was to perform penance after committing an environmental offense, such as destroying plants or animals. Kautilaya Arthashastra helps in determining the ecological sustainability of the Mauryan period. The Arthashastra gives a demonstration of Kautilya's concern and perception of living animals, both domestic and wild. Penalties and strict punishments were imposed for injuring these living creatures.

9.3.2 WATER

Water from time immemorial has always been an important natural resource needed for sustaining life. Ancient Indian tradition places great emphasis on maintaining and nurturing water bodies using all available water purification techniques (e.g., jala-kataka-renu powder) and also implementing rainwater harvesting methods.

However, most of the ancient hydrologic knowledge remained hidden from the world at large until recent times. The Vedas, particularly, the Rigveda, Yajurveda, and Atharvaveda, have many references to the water cycle and associated processes, including water quality, hydraulic machines, hydro-structures, and nature-based solutions for water management. The Harappan civilization epitomizes the level of development of water sciences in ancient India that includes the construction of sophisticated hydraulic structures, wastewater disposal systems based on centralized and decentralized concepts, and methods for wastewater treatment. The construction of step wells to bridge the gap in water supply and demand during periods of scanty monsoon, reservoirs to regulate water flow and prevent flooding, and tanks for storing harvested rainwater are a few examples that demonstrate the high level of expertise in water conservation during that time period. Vedic scriptures carefully drafted the management of water resources by taking water footprints, exploring potential hidden water sources, and ensuring proper drainage systems. Caraka Samhita addressed possible water contamination issues by doing basic tests in terms of color, odor, taste, and touch. The Mauryan Empire is credited as the first "hydraulic civilization" and is characterized by the construction of dams with spillways, reservoirs, and channels equipped with spillways (Pynes and Ahars). They also had an understanding of water balance, the development of water pricing systems, the measurement of rainfall, and knowledge of the various hydrological processes [21].

9.3.3 AIR

In the absence of deforestation and industrialization, air pollution was not a big concern in ancient times. But still, sadhus used to perform Yagya which was the scientific process of sublimation and transformation of healthy constituents of herbal medicines into the vapor phase and their expansion and dissipation in the surrounding area. It also helped in maintaining the balance of oxygen, carbon dioxide, and other harmful growth of poisonous gases and radiations [22].

9.3.4 LAND

The people of the Indus Valley Civilization lived in multistoried buildings constructed of baked bricks and were known for their strong sense of civic responsibility. There were large roads paved with baked bricks, well-laid drainages, aqueducts, private and public baths, and other sanitary features. These aspects are mostly addressed during the Mauryan period. As per Kautilya's Arthshashtra, residential buildings, roads, commercials, cremation grounds, etc. ought to be constructed in such a way that they cannot harm the ecology and don't hurt biodiversity. Every house should have a proper system for controlling fire, sewage removal, and waste disposal. Any sort of violation of these rules was legally liable to penalty or punishment. He also prepared a disaster management system to tackle natural calamities [23].

9.3.5 HEALTH

"Ayurveda," meaning the science of life, is highly systematized and organized medical knowledge evolved in ancient India (between 2500 BC and 1000 BC). Caraka and Susruta, the two ancient masters of Indian medicine, observed that health depends on heredity, geographical environment, climate, water supply, quality of air, time, and seasonal variation. Ayurveda holds a huge knowledge of green food and the ability to cure diseases which are mentioned in Carakasamhita, especially in Suthrasthana. Medicines were chiefly derived from plants, although a few ingredients originated from animals. Preparations of medicines involved the collection of the ingredients, their purification, extraction of their essences, and compounding of these extracts using processes like grinding, pasting, and maceration. Processes like dissolution, distillation, sublimation, precipitation, combustion, dilution, and decocting were carried out in these preparations. Mercury and gold were also used in a number of drugs. The "Sanjivani buti" mentioned in Ramayana which helped Laxman gain consciousness suggests the effectiveness of drugs in ancient India. Vrukshayurveda is a branch of Ayurvedam that describes the science of medicinal system for the benefit of life which are still being practiced in modern India [24–26].

9.3.6 INDUSTRY

Ancient Indian history does not represent extensive industrialization; it was limited to fulfilling basic material needs. It mostly consisted of construction; brick manufacturing; medicines, cosmetics, and perfumes; glass making; pottery; paper; dye; leather; clothing; jewellery, art, and artifacts; and also alcoholic liquors. Our ancestors revered seers who had a profound belief in various cultural texts that showed a thorough understanding of the negative impacts of climate change, such as irregular weather patterns, water pollution, imbalances in the ecosystem, and environmental degradation. The people were cautious enough not to break the divine rules and disrupt the weather cycle, rainfall phenomena, hydrologic cycle, and life cycle [27].

9.4 CHALLENGES AND POSSIBLE SOLUTIONS

Although green chemistry has the power to bring about radical change by offering "greener" reagents and alternatives and more benign routes to synthetic methodologies, it is yet to get the awareness it deserves as some industrialists still consider it as a fashion and not a necessity. The malevolence of chemical technology was observed in a number of cases, but it was the Bhopal gas tragedy that shook the world and made policymakers and chemical practitioners arrive at a consensus to avoid such a disaster in the future. In order to ensure global environmental protection while keeping scientific and economic development at the forefront, policymakers should understand the role of "green" science and technology and make pollution prevention the main aim rather than pollution control. It is necessary to create, enforce, and regularly monitor strict rules and regulations.

It is also necessary to promote the awareness of green chemistry at the local level through the involvement of not only the academia or academic intelligentsia but also the science and technology agencies and the administrators. Academic institutions must devise a curriculum so as to make future generations realize the pertinence of green chemistry. Furthermore, funding from research and development and the science and technology agencies must be restricted to projects following green principles. A collective effort from all the stakeholders will make green chemistry a synchronized movement and bring about a green revolution in chemistry and chemical technology [28].

REFERENCES

1. Kidwai, M. (2001) 'Green Chemistry in India', *Pure and Applied Chemistry*, 73(8), 1261–1263.
2. Anastas, P.T., Williamson, T.C. (1996) 'Green Chemistry: Designing Chemistry for the Environment', in ACS Symposium Series, Washington, 626.

3. Ray, P.C. (1909) *A History of Hindu Chemistry*, Vols. I & II. London: Williams and Norgate.

4. Verma, S. (2018) 'Chemistry in Ancient India', https://www.pgurus.com/chemistry-in-ancient-india/.

5. Ali, M. (1993) 'A Brief History of Indian Alchemy Covering Pre-Vedic to Vedic and Ayurvedic Period (Circa 400 B.C.-800 A.D.)', *Bulletin of the Indian Institute of History of Medicine (Hyderabad)*, 23(2), 151–166.

6. Deshpande, V. (1998) 'History of Chemistry and Alchemy in India from Pre-historic to Pre-modern Times', in A. Rahman (Ed.), *History of Indian Science and Technology and Culture AD 1000–1800*. Delhi: Oxford.

7. Subbarayappa, B.V. (1999) 'Indian Alchemy: Its Origin and Ramifications', in *Chemistry and Chemical Techniques in India*. Delhi: Centre for Studies in Civilisations.

8. Habib, I. (2000) 'Joseph Needham and the History of Indian Technology', *Indian Journal of History of Science*, 35(3), 245–274.

9. Pathak, J. (2016) 'Chemistry in Ancient India', *IRJMST*, 7(3), 157–167.

10. Caraka, Sutrasthaman, Engineering Transactions P.V. Sharma, Text with English translation, Chaukhamba Orientalia, Varanasi, sixth edition, 2001, ch. I, 64-66.

11. Susruta Samhita, Sutrasthanam, Sutra, Ch XLII, 2. (Note: For Susruta Samhita the work followed is Kaviraj Kunjalal Bhisagratna, Text with English Translation, in 3 volumes, Chowkhamba Sanskrit Series Office, Varanasi, first edition 1998).

12. White, D. (1996) *The Alchemical Body: Siddha Traditions n Medieval India*, University of Chicago Press, 80.

13. Dasgupta, N. (2009) 'Rasa, Rasayana, Rasatantra: Exploring Concepts and Practices', *An Ancient Indian System of Rasayana*.

14. de Marco, B.A., Rechelo, B.S., Tótoli, E.G., Kogawa, A.C., Salgado, H.R.N. (2019) 'Evolution of Green Chemistry and Its Multidimensional Impacts: A Review', *Saudi Pharmaceutical Journal*, 27(1), 1–8.

15. Yamamoto, S. (1998) 'Contribution of Buddhism to Environmental Thoughts', *The Journal of Oriental Studies*, 8, 144–173.

16. Dixit, S. (2018) 'Environmental Sustainability Lessons from Ancient India', *Keemat*: September – October 2018, 9–12.

17. Ahmad, F. (2020) 'Origin and Growth of Environmental Law in India', *Journal of the Indian Law Institute*, 43(3), 2001, 358–387.

18. Thakur, B. (2019) 'History of Environmental Conservation (Ancient and Medieval Periods)', *International Journal of Multidisciplinary Research Review*, 4, 1072–1077.

19. Manjrekar, S.M. (2017) 'Study of Sustainability Practices Ingrained in Indian Culture', *5th International Conference on Civil, Architecture, Environment and Waste Management (CAEWM-17)*, 27–30.

20. Patra, B. (2016) 'Environment in Early India: A Historical Perspective', *Environment: Traditional & Scientific Research*, 1(1), 47–48.

21. Singh, P. (2020) 'Hydrology and Water Resources Management in Ancient India', *Hydrology and Earth System Sciences*, 24(10), 4691–4707.

22. http://literature.awgp.org/akhandjyoti/2006/Jan_Feb/v1.VedicSolutionPureAir.

23. Sharma, R. (2014) 'Ecological Sustainability in India through the Ages', *International Research Journal of Environmental Sciences*, 3(1), 70–73.

24. Purwar, C. (2016) 'Significant Contribution of Chemistry in Ancient Indian Science and Technology, *International Journal of Development Research*, 6(12), 10784–10788.

25. Wewalwala, S.L. (2011) 'Ayunreda and Sustainable Development', *Economic Review*, 2011, 29–32.

26. Harbans Singh Puri (2002) 'Rasayana: Ayurvedic Herbs for Longevity and Rejuvenation', General Series on Traditional Herbal Medicines for Modern Times, Gen.ed., Dr Roland Hardman, CRC Press, Taylor and Francis, London, New York, 2002, vol 2, viii.
27. Ray, P. (1948) 'Chemistry in Ancient India', *Chemistry in Education*, 25(6), 327.
28. Bora, U. (2002) 'Green Chemistry in Indian Context – Challenges, Mandates and Chances of Success', *Current Science*, 82(12), 1427–1436.

Index